Rad Bugs

Brain Worms, Ticks that Cure Cancer, and Other Tiny Marvels

Karyn Light-Gibson

Microcosm Publishing
Portland, OR | Cleveland, OH

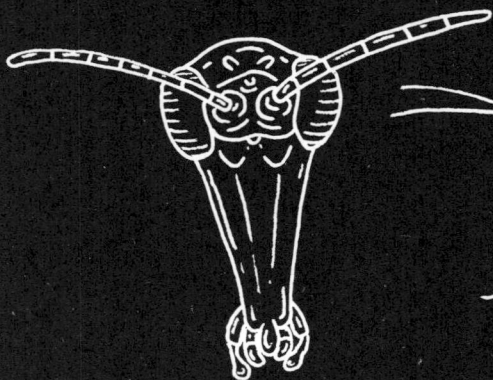

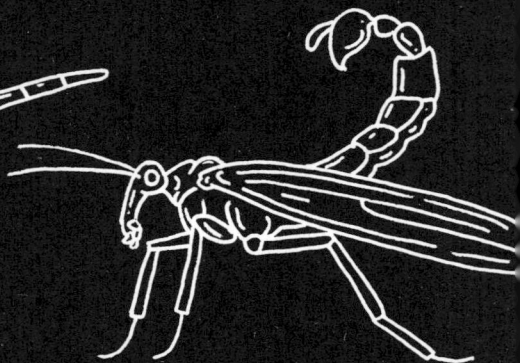

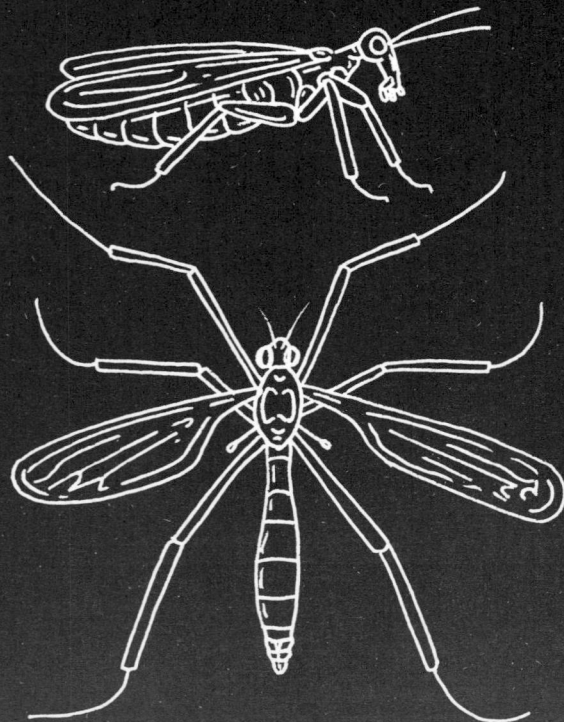

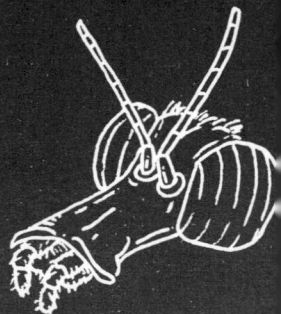

Rad Bugs

Karyn Light-Gibson

Microcosm Publishing
Portland, OR | Cleveland, OH

Rad Bugs: Brain Worms, Ticks that Cure Cancer, and Other Tiny Marvels

© Karyn Light-Gibson, 2026

This edition © Microcosm Publishing, 2026
First Edition, 3000 copies
Paperback ISBN 9781648415951
Ebook ISBN M69518

This is Microcosm # 1135
Edited by Ivy Zeller
Designed by Joe Biel and Sarah Koch
Illustrated by Karyn Light-Gibson

For a catalog, write or visit:
Microcosm Publishing
2752 N Williams Ave.
Portland, OR 97227

All the news from the misfits in print at Microcosm.Pub/Newsletter
Get more copies of this book at Microcosm.Pub/RadBugs
EU Safety Information: microcosmpublishing.com/gpsr

To join the ranks of high-class stores that feature Microcosm titles, talk to your rep: In the U.S. COMO (Atlantic), ABRAHAM (Midwest), BOB BARNETT (Texas, Oklahoma, Arkansas, Louisiana), IMPRINT (Pacific), TURNAROUND (UK), UTP/MANDA (Canada), NEWSOUTH (Australia/New Zealand), Observatoire (Africa, Europe), IPR (Middle East), APD (Asia), HarperCollins (India), and FAIRE in the gift trade.

Did you know that you can buy our books directly from us at sliding scale rates? Support a small, independent publisher and pay less than Amazon's price at www.Microcosm.Pub.

Global labor conditions are bad, and our roots in industrial Cleveland in the '70s and '80s made us appreciate the need to treat workers right. Therefore, our books are MADE IN THE USA.

Microcosm's workers and authors are paid solely from book sales. If you downloaded this book from some sketchy part of the Internet or picked up what appears to be a bootleg, please support our hardworking team by purchasing a copy directly from us and encouraging your communities to do the same. Paying for our books and zines helps us publish work that's far better than anything AI can come up with. Additionally, a 2025 MIT study revealed that AI inhibits humanity's critical thinking ability. Since critical thinking is one of our core values, we prohibit any use of our books to "train" generative artificial "intelligence" (AI) technologies, because seriously, WTF?

MICROCOSM PUBLISHING

MICROCOSM PUBLISHING is Portland's most diversified publishing house and distributor, with a focus on the colorful, authentic, and empowering. Our books and zines have put your power in your hands since 1996, equipping readers to make positive changes in their lives and in the world around them. Microcosm emphasizes skill-building, showing hidden histories, and fostering creativity through challenging conventional publishing wisdom with books and bookettes about DIY skills, food, bicycling, gender, self-care, and social justice. What was once a distro and record label started by Joe Biel in a drafty bedroom was determined to be *Publishers Weekly*'s fastest-growing publisher of 2022 and #3 in 2023, and is now among the oldest independent publishing houses in Portland, OR, and Cleveland, OH. Biel is also the winner of PubWest's Innovator Award in 2024. We are a politically moderate, centrist publisher in a world that has inched to the right for the past 80 years.

Contents

INTRODUCTION

*W*ell, I'm back because more cool bugs need the spotlight.

Maybe you picked up this book because you read *Bug Life*, my first book, and couldn't wait to get more bug facts. Or maybe you're reading this with no other knowledge of bugs. Regardless of how you show up to this book, I hope you enjoy all the fun bug information that you can whip out as trivia at parties, use to help your grandma get over her disgust of slugs, or finally decide on that dope centipede tattoo you've been wanting to get.

The first book, *Bug Life*, included only insects. While they are super important and interesting, the title *bug* applies to so many other insects and non-insects. Entomologists state that "true bugs" are only found in the order Hemiptera. But in common language, really anything without a backbone (also called an *invertebrate*) gets that title, especially if it is terrestrial (lives on land, like centipedes and slugs) or in freshwater (like mussels and crawfish). You only need to go to the r/whatsthisbug or r/insects subreddits to realize how broad the category *bug* is.

My fascination with bugs has been strong since I was a kid. Being the little weirdo that I was, I loved to "help" the spiders, who lived in our bushes: I would pick up ants, drop them in the spiders' webs, and watch them run out to wrap the ants up. I realize as I write how macabre that is, but I was a strange child.

And I thought I was helping. Good intentions, misplaced practice. Spiders don't need our help catching food; they just need our help making sure we don't completely destroy the planet and their food sources (other bugs). Currently, the biggest threat to bugs is . . . us.

As I grew and learned more, I realized there are a lot of misconceptions and myths surrounding bugs. People frequently only see them as pests, but these creatures are so much more. I, even as a lover of bugs, have realized how easy it is to believe incorrect information through media and word of mouth. The disconnect between the assumptions and the truth are pretty stark.

After studying language, culture, and biology in grad school, I wanted to spend my time creating materials like this book to help people understand the great importance of invertebrates to our world. And how we need them more than we realize.

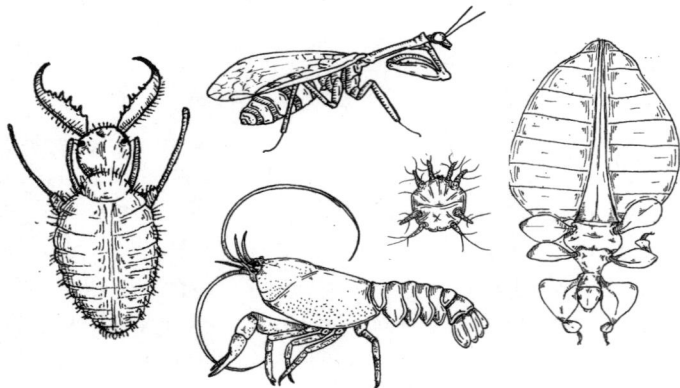

Bugs in various forms have survived numerous extinctions. In fact, scorpions go as far back as five hundred million years. And silverfish have been around and appear unchanged (at least externally) over the last four hundred million years.

Humans may give them a run for their money, however. The Anthropocene, which we are currently in, is the sixth extinction event. This extinction event is human-caused, not something like an asteroid or a large volcanic explosion. Habitat destruction, global warming, pollution, and invasive species are all causing serious issues for bug species.[1]

---

1 Those things are killing us as well. All life loses because of them.

DEFINING TERMS

So how do I define *bug* in this book? I'll be using it to refer to most invertebrates—those without backbones. You can identify arthropods (another word for the backbone-less) by how many body segments, limbs, eyes, and mouthparts they have. For example, one quick way to determine the type of bug a spineless creature is? Count their legs:

- Six legs: it's an insect
- Eight legs: it's an arachnid
- More than eight legs/no legs: probably a bug, but not an insect or arachnid

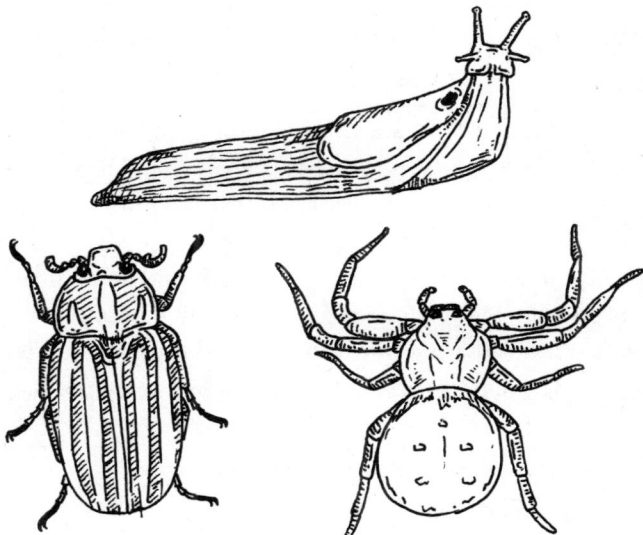

Insect (left), arachnid (right), and slug (top)

Many of the bugs in this book fall into the eight legs or more category.

However, scientific classification, also called *taxonomy*, can be confusing, so I'm defining some important terms for the journey ahead.

Here's one popular elementary-school mnemonic device to remember the different parts of this system:

Kangaroos **P**lay **C**ards **O**ut **F**ront, **G**o **S**ee

Kingdom **P**hylum **C**lass **O**rder **F**amily **G**enus **S**pecies.

All life (that we know of) is placed into these categories created by western scientists, with *kingdom* as the broadest category and *species* as the most specific. Humans and insects are part of the kingdom Animalia, but we split off after that. Phylum Arthropoda contains all invertebrates (and thus, all bugs) on earth.

	HUMANS	BUGS
KINGDOM	ANIMALIA	ANIMALIA
PHYLUM	CHORDATA	ARTHROPODA

Taxonomy is complicated and ever-changing with new discoveries. In this book, the chapters may have information about different *suborders*. A suborder is a way to classify creatures within an order into smaller, more closely related groups, before you get down to the level of families. These levels aren't easy to identify, and sometimes they shift. Arthropods are shifting especially frequently, as more species are discovered. Just know that *suborder* comes between *order* and *family*.

Lastly, here is a quick glossary of bug words I will use repeatedly in the pages to come:

- Arthropod: any animal without a backbone
- Insect: an arthropod with six legs and three parts
- Arachnid: an arthropod with eight legs and two body parts (mostly)
- Thorax: the middle of three parts on an insect
- Abdomen: the end (butt) of the three parts of an insect
- Tarsi: the equivalent of toes on bugs
- Exoskeleton: the hard outer shell of arthropods

- Exeuvia: the shed of an insect between some life stages, like the shell of a cicada
- Simple metamorphosis: the bug looks the same throughout its life, just in different sizes
- Complete metamorphosis: when the young and the adults of a species look very different
- Larva: the first form for insects with complete metamorphosis
- Pupae: for insects with complete metamorphosis, the form between the young and adult stage—kind of an adolescence, in which the insect does not usually feed or move around much
- Nymph: the young form of insects with simple metamorphosis
- Instar: a life stage of insects who do not go through full metamorphosis; often looks like smaller versions of their adult selves

QUICK FACTS: BUGS

- Most bugs are not pests. Only about one to three percent of all known species are what we consider human pests.
- Humans cannot live without bugs. We'd be dead very soon—probably about six months—after they disappeared.
- Ninety-seven percent of all life on earth is invertebrate.
- Bugs need our help to survive. Even though they've survived numerous extinctions, their numbers are dwindling in theAnthropocene (the time that human activity has been the dominant influence on climate and the environment).
- Most representations in popular media (films like *Arachnophobia* and *The Bay*) are highly inaccurate or purposefully play on innate fears about bugs.

Have we always had such a fear of bugs though? Surely, since we've lived with them since the existence of people, there must be something to our current level of dislike for bugs.

Yuya Fukano and Masashi Soga, two researchers from Japan, came up with the idea of the urbanization/disgust hypothesis in 2021. This hypothesis states that the issue with bugs comes from an increase in urbanization and a general lack of knowledge of bugs. In urban spaces, we are more likely to see pest species. And since we don't know much about them, our disgust increases exponentially.

This book tries to bridge some of that knowledge gap and help everyone appreciate and understand these creatures better.

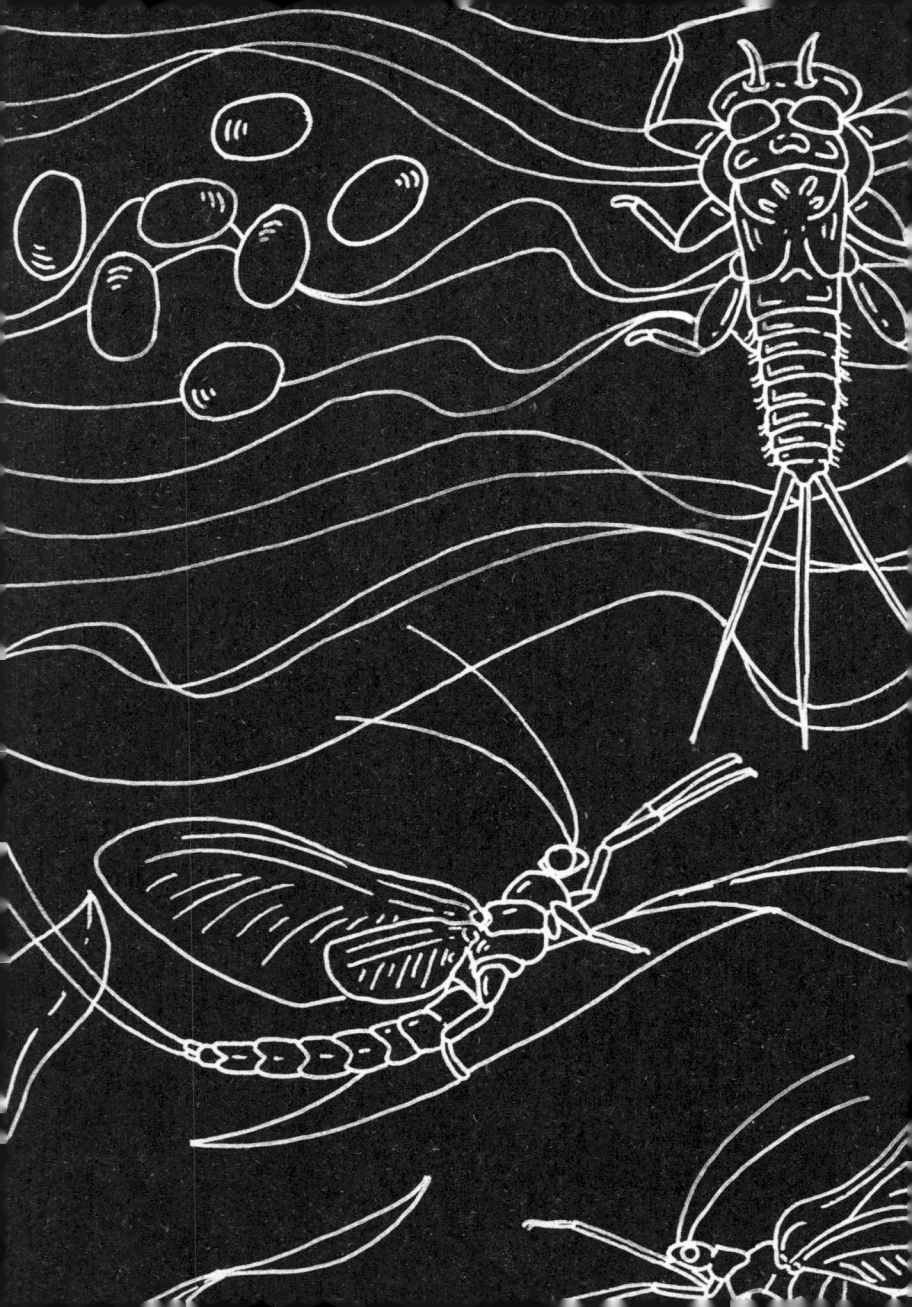

Part I: The Less-Often-Seen Ones—Some Insect Orders

Some bug orders are huge; however, some are fairly small (at least compared to orders, like beetles). In this section, I will highlight a few of them that you may see, especially in urban spaces. Just because they are smaller orders or less-often-seen does not make them any less important or impactful.

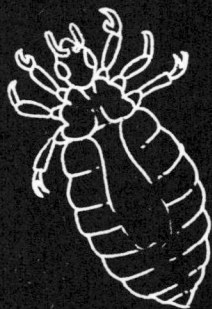

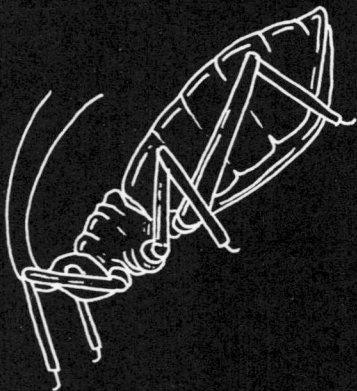

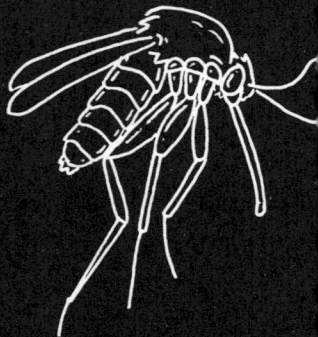

CHAPTER 1: FLEAS (ORDER SIPHONAPTERA)

*M*any people associate fleas with dirty environments—which isn't always accurate because these creatures can show up anywhere. For example, a friend of mine, whose cat is indoors and never leaves the apartment, was scratching furiously. And the vet said, "Fleas." My friend was shocked since they had not seen them on the cat or anywhere else. The vet mentioned that fleas can come from anywhere so you never know where the infestation started (similar to bedbugs). Even though they can show up in any space, fleas can flourish in messy environments.

Flea fossils are few and far between. From what is available, scientists believe that fleas have been around since the Jurassic period but don't know which other insects they evolved from or are closely related to. Their best guess is that they evolved from flies, but that is still debated.

For sustenance, fleas mostly rely on mammalian hosts, but some of them live on birds.

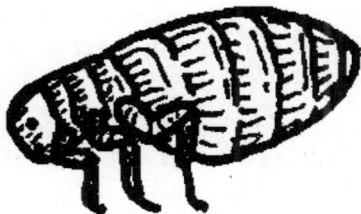
Boing boing!

The adults are pretty flat from the side with impressive jumping legs. They can't jump as well as some of their hemipteran friends, but they are able to jump seven inches high and thirteen inches horizontally, which is pretty impressive for an animal that is about 1/16 of an inch long.[2] They also can jump two hundred times their body length, which would be like a human adult jumping a quarter mile.

2 Michigan State University Extension, "Fleas," 2007, https://www.canr.msu.edu/ipm/uploads/files/Fleas.pdf.

How Fleas Shape Our World

Fleas like to live on animals, including domesticated animals. If they are not addressed quickly, they can spread to humans and human dwellings. Once their numbers are large enough that the animal is too small to contain them, they find other places to live. But if caught early enough, you don't have to worry about an infestation. If their numbers continue to grow, they will bite humans. Their bites can be maddeningly itchy and have the potential to be dangerous: fleas are to blame for the spread of the bubonic plague and can also spread typhus.

Because of flea-borne illnesses, these tiny critters have had a major impact on human history. The term *quarantine* comes from the forty-day timeframe infected sailors had to spend on ships before they could step on land: *quarant* is French for "forty."

The Black Death, a bubonic plague pandemic, killed upwards of half of all Europeans between 1346 and 1353—one of the deadliest recorded pandemics. It not only affected Europe but also Asian and African countries as well. The number of deaths completely altered the economy and culture of those places hit the hardest, a major flea-borne tragedy.

Scientists still debate where the infected fleas/rats came from and how it spread the way it did. A large factor was probably the lack of cleanliness in Europe at that time: lots of human and animal waste littering streets and doorsteps, wild and farmed animals roaming the streets, and unawareness of germs in general.

Believe it or not, the bubonic plague still exists. Madagascar experienced an outbreak in 2014 which killed over one hundred people and infected thousands. In the US, around seven cases of the plague are reported every year, most often occurring in New Mexico, Arizona, and Colorado. Antibiotics, which were not available in the fourteenth century, greatly reduced the mortality rate (about eleven percent—the seasonal flu is around four percent).[3]

---

3 Center for Disease Control and Prevention, "Frequently Asked Questions," Plague Home, November 15, 2021, https://www.cdc.gov/plague/faq/index.html.

How Humans Can Interact with Fleas

Most of the time, fleas do not carry these diseases. You still don't want them hanging around your house, as they are annoying (and it's better to be safe than sorry).

Once humans discovered germs and established ideas of cleanliness, fleas became associated with only those who were "unclean," however the elite chose to define the term. Fleas still live in human dwellings, but similar to other hard-to-eradicate bugs, the choice to get rid of them may be outside the reach of those that need it most.

Things like time, money, and shame have a great effect on whether or not people can remove the burden from their homes. In 2022, San Francisco and Los Angeles had numerous cases of typhus, which is a disease rarely seen in cities in the US. There had not been a death from typhus in LA since 1993, and in 2022, there were 3 deaths out of 171 confirmed cases. Scientists acknowledged that income, healthcare access, and insurance coverage most likely played a role in those who eventually needed treatment for typhus.[4]

Representatives in LA stated they believed the infections were linked to the lack of care given towards the unhoused. Certain laws went into place within the last few years that have increased numbers of displaced residents. Health hazards, such as human waste, have become a serious issue for the people who find themselves in unhoused situations. Instead of helping these people, the city has continued to avoid the problem which, if tied to flea-carried typhus, will only continue to create larger issues.

Unfortunately, humans also decided to harness flea-borne illness for their own benefit, to the detriment of others. Fleas have been weaponized for war with devastating results. Records from the fourteenth and eighteenth centuries show that armies intentionally threw plague bodies over walls to "smoke" out their enemies or cause

4   Lucas S. Blanton, "Rickettsia typhi in Southern California: A Growing Flea-Borne Threat," The American Journal of Tropical Medicine and Hygiene 110 no. 1 (2023), https://doi.org/10.4269/ajtmh.23-0742.

bouts of the plague within cities. Japan dropped fleas with the plague from planes onto parts of China, causing small pockets of infections within the country. The US and the USSR ramped up testing of insects as biological vectors during the Cold War.

The US conducted a test run called Operation Big Itch in 1954 to determine the success of dropping plague-fleas in inhabited areas. In the test operation, they used non-plague fleas. They wanted to see if the fleas could survive being dropped in a bomb from a plane. The first couple of tests went hilariously wrong, including an incident in which the fleas were opened in the plane and bit the pilot and others on board.

Their future tests proved that fleas would survive the drop. They were never able to try it in war, at least with fleas. In 1972, the use of insects infected with biological vectors in war was banned. A committee of experts determined that if the plague was released on a populace that was not ready for it, it could spread to ninety percent of the population within a month, killing sixty to seventy percent of people.[5] Because of these numbers, the Toxic Weapons Convention decided to ban bug vectors across the world.

How Human Culture Is Shaped by Fleas

John Donne has a famous poem called "The Flea" that uses the animal to reference seduction. (He's just trying to get a lady to bone him.) It didn't hold up well, as it comes off as pretty misogynistic.

In more recent cultural events, the bassist for the Red Hot Chili Peppers is called Flea. He got that nickname in school because of his inability to sit still, and he's carried the name with him ever since.

Especially popular during the 1800s and early 1900s, flea circuses were a simple form of entertainment. Some did not contain fleas and were simply pulley systems and independently moving carts. Some did contain actual fleas, which were then glued to small instruments or

5 David. T Dennis, "Plague as a Biological Weapon," Bioterrorism and Infectious Agents: A New Dilemma for the 21st Century, (2009), https://doi.org/10.1007/978-1-4419-1266-4_2.

bicycles. Most circuses used mole fleas, as they are larger and don't jump nearly as far as some other species. Sometimes heat was applied under the fleas to get them to move by trying to wiggle free from their stuck position. While fleas are not beloved, this treatment does seem harsh.

Cute? Maybe. Cruel? Definitely, even for something so unloved as a flea.

"Flea markets" are common in towns and cities. Sometimes called bazaars or indoor swap meets, a flea market is traditionally a non-permanent used goods sale, made up of mostly clothes and furniture. There is some debate as to where the term originated from, but most believe it came from the French phrase *marché aux puces*, which translates literally to "market of fleas."

Fleas are often viewed as pests, but they thrive in a world of loss: They have found ways to survive, where others couldn't. Fleas have adapted to be successful parasites and we just happen to be the ones on their menu. Everyone has their ecological niche and fleas have theirs, even if we don't like how much we are unable to control them.

QUICK FACTS: FLEAS

- They have been weaponized by humans in war with devastating results.
- The plague, a flea-borne illness, isn't completely eradicated, and there are cases every year.
- Fleas don't only show up in dirty environments, but they do thrive there.
- While they may not jump as well as some hemipterans, they can jump seven inches high and thirteen inches across.
- They can show up out of nowhere, and it is hard to trace where they came from.

CHAPTER 2: LICE (ORDER PSOCODEA)

*W*hen I was younger, my four cousins came over for the week and I found a louse in their hair about thirty minutes after they'd arrived. We went into full lockdown mode. My mom bought the shampoo and combs but my aunt just wanted to shave all of their heads. At this, the cousins were devastated. My aunt swore that shaving their heads would be the best way to eradicate the lice.

Unfortunately, the general consensus is that shaving your head will not get rid of lice. Lice can hold on to very short hairs and their eggs can be laid very close to the scalp. Just shaving does not help but may make combing for nits easier. So, don't shave your head, your kids' head, or your friends' head and think that you've taken care of the issue.

How Lice Shape Our World

Let's talk about the harmless side of the order Psocodea. Book lice and bark lice are the less-often-encountered cousins of head lice. They don't have the same pest status and are frequently unnoticed by people. It helps that they are very small (like their cousins) and not often found in human dwellings.

Book lice aren't pesty in the same way as parasitic lice, but they can cause issues in food storage facilities. They reproduce quickly and can cause foods to become inedible. Their name comes from their interest in eating books. Unlike carpet beetle larvae who love book glue, book lice will eat mold and mildew that appears on the pages or the spine of a given book, in addition to the glue that holds it together.

They are found more often in humid spaces and are less likely to be found in houses that have central heating and cooling (as it makes the air drier). Sometimes, book lice are mistaken for bedbugs. Book lice are very small (around one millimeter) and have very long antennae, as well as very apparent, black eyes. (Bedbugs can be small in earlier instars but have short antennae and a rounder body.) They are the original "book worm," even though many different insects can eat books. Keeping your books (and grains) free of moisture is the way to stop them from coming inside.

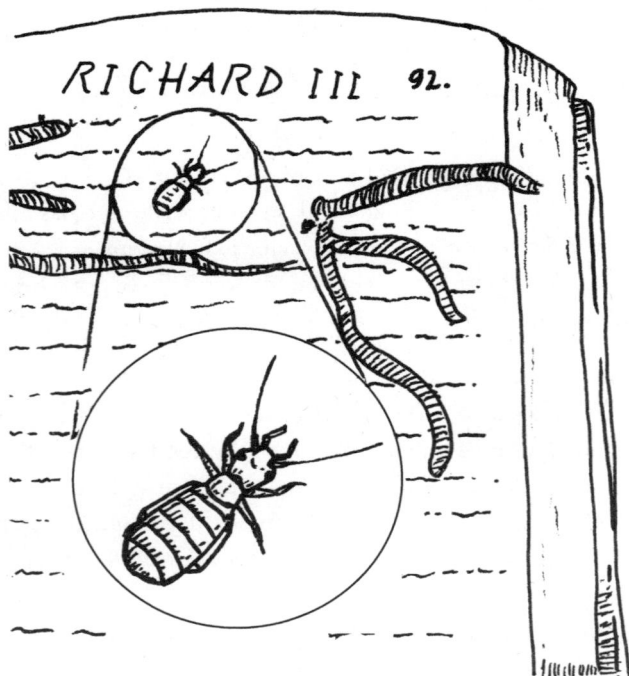

Just getting ready for a read and a snack.

Bark lice get their name from their preferred habitat: trees. They live on trees eating organic material, lichen, mold, fungus—all the things that grow on trees in humid environments. They are not detrimental to the trees they live on and should not be concerning to people who see them. One species moves in interesting patterns together, almost like a herd of cattle moving across a tree. Another species creates a web-like structure on the trees they live on. The structure does not damage the tree and does not need to be removed.

How Humans Can Interact with Lice

The order Psocodea includes bark lice, book lice, and parasitic lice. Many of the species are not associated with humans or human bodies. Those that are thankfully are easier to get rid of than some pests (bedbugs, fleas, etc.). Some species, like the body and head louse, have been known to spread typhus, but that only happens in larger numbers and unsanitary conditions. Most of the interactions with head lice are very unlikely to cause any serious issues, more just annoyances. The phrase "feeling lousy" comes from the word *louse*, which is the name for an individual bug—*lice* is for more than one. The phrase means "to feel bad or disgusting." Many people find lice disgusting since they live on humans and eat their blood. Fair enough.

So few of them fall into that category, however. Many species in the order do not rely on human blood (or any blood for that matter). About thirty-four percent of bird species and twenty-seven percent of mammalian species have a few different types that live on them.[6] Frequently, lice do not want to kill or weaken their host, since they rely on them. Many birds live with lice throughout their life in small enough numbers that keep both the lice and bird healthy. In larger numbers, weakness and fatigue lead to other issues that can cause serious problems with the host (similar to fleas). Other insects in this order, like bark lice and book lice, eat organic matter, not blood: fungus and book-binding glue respectively.

Parasitic lice are perhaps the most well-known. Only three species out of the recorded 11,000 feed on humans: body lice, head lice, and pubic lice. These species cannot be spread by your pets. Head lice is commonly found making its rounds in schools on children. Most people will experience this type of lice once in their life. Often spread at schools, lice can easily spread from head to head through hats,

6 Vincent H. Resh and Ring T. Cardé, eds., Encyclopedia of Insects, 2nd edition (Academic Press, 2009).

sleeping close together, or other activities that put people in close proximity. They do not have wings so they must crawl from host to host. Similar to bedbugs, they have figured out a way to thrive without needing wings.

The lice themselves are visible but their little eggs, called nits, are particularly difficult to see. The word *nitpicking* is a reference to picking out the tiny eggs. It was first used in print on December 21, 1951 in *Charleston Daily Mail*: "Sen. Johnson is encouraged to proceed with his nitpicking."

Parents may spend a lot of time combing nits out of hair. Up until the nit comb was created, humans had to pick off lice and nits by hand. There are numerous paintings and drawings from the 1600s showing parents picking lice off of their children. Similar to apes and other primates, humans would need to groom those with lice as the only way to remove them. There are over the counter shampoos that people can use to get rid of lice, but doctors are beginning to see resistances in some species. Doctors recommend a topical cream or oral ivermectin to make sure that the lice are killed. Using store-bought shampoos may not get rid of their presence (similar to DIY bedbug treatment). Talking to a professional is recommended before trying anything on your own.

Pubic lice, also called crab lice, are found most often in pubic areas but can also be found in other hairy parts of the body, like eyelashes. They are called crabs because the lice themselves are crab-shaped: stout bodies with large claws on their second and third pairs of legs. They are most commonly spread through sexual interactions but may also be spread by towels or bedsheets. A Bloomberg article from 2013 states that there is anecdotal evidence of a decrease in pubic lice numbers linked to removal of pubic hair (more specifically the popularity of Brazilian waxes).[7] While this has not been confirmed, researchers believe there may be a connection between the two.

7 Jason Gale and Shannon Pettypiece, "Brazilian Bikini Waxes Make Crab Lice Endangered Species," Bloomberg, January 13, 2013, https://www.bloomberg.com/news/articles/2013-01-13/brazilian-bikini-waxes-make-crab-lice-endangered-species-health.

When pubic lice are present, they cause the area they are living in to itch. Both pubic lice and head lice can happen to anyone at any time and have nothing to do with cleanliness. In fact, they prefer clean places. Both species only live on humans, so they won't show up on your cat (if your cat has lice, you probably can't get that species from them).

Body lice are the only psocodean species that has connections to lack of cleanliness, but the lice do not exclusively live on those who don't wash. They may have just started somewhere unclean: dirty bed sheets, towels, or clothing. If you go to a hotel where they haven't changed the sheets since the last guest and that guest had body lice and you are now sleeping in that bed, you could get lice from that previous guest. Eggs can be killed when exposed to high heat, so washing and drying clothing and bedding above 130 degrees should take care of them. Doctors prescribe a cream that kills them. Between that and drying your things on high heat, body lice can clear up pretty quickly.

Unfortunately, the situations in which these lice have the highest possibility of causing problems are when things like hot laundry and medicine may not be readily available: troops in war, settlements of unhoused people or shelters, and people displaced by natural/other disasters.

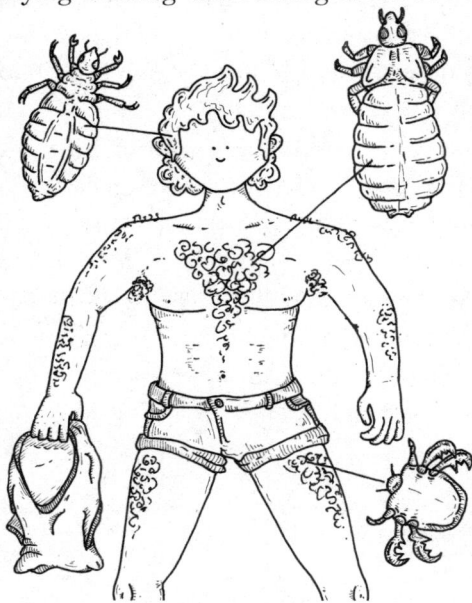

This guy's got 'em all!

How Human Culture Is Shaped by Lice

Body lice are the only human lice species known to carry disease, but this is only likely to happen in unsanitary, crowded conditions. Issues related to lice-carried typhus have been seen in soldiers during war, especially when in trenches during World War I and II. Lice also tore through Napoleon's army by spreading disease.[8]

When his soldiers were retreating from Russia, he lost a lot of men. In the early 2000s, researchers tested a mass grave to see if the assumption that lice-borne infections caused significant deaths in their ranks. They concluded that upwards of one-third of the troops may have died from typhus that had been spread by lice. That means that 125,000 men may have died from the illness carried by a small animal and not from fighting in the war. Without access to clean clothes or beds, the men experienced the lice moving freely between them. Typhus is only a secondary effect of the lice, however, as they do not mean to spread it. Their only desire is to eat. They just also happen to carry diseases that can be detrimental to their host.

Some religious figures chose to keep lice as a sign of religious devotion. Historian Katherine Harvey compares it to flagellation: a way to show the depth of their faith.[9] They would allow themselves to get lice and not make any attempts to get rid of them. The pain and suffering caused by the lice was a way to lead them to spiritual purity.

The term *louse* has carried a negative connotation for some time. In Shakespeare's 1602 play *Troilus and Cressida*, one character implies that another character is like a louse on a leper: a double-burn.[10]

---

8 Didier Raoult, Olivier Dutour, Linda Houhamdi, Rimantas Jankauskas, Pierre-Edouard Fournier, Yann Ardagna, Michel Drancourt, Michel Signoli, Vu Dang La, Yves Macia, and Gérard Aboudharam, "Evidence for Louse-Transmitted Diseases in Soldiers of Napoleon's Grand Army in Vilnius," The Journal of Infectious Diseases 193, no. 1 (2006), https://doi.org/10.1086/498534.

9 Katherine Harvey, "Medieval Parasites," Aeon, April 9, 2019, https://aeon.co/essays/medieval-people-were-surprisingly-clean-apart-from-the-clergy.

10 NOTE: Shakespeare really had some great insect-related content. Check out the article "Shakespearean Entomology" by David Miller at the New Zealand Electronic Text Collection.

In 2023, entomologists extracted DNA for lice that had died long ago. In their DNA analysis, they found two distinct genetic clusters. This means that the lice had come from two different places. The team wanted to showcase how important lice have been in human history and how we could use them to better track movement across continents. One entomologist stated that lice are "really like a little tape recorder that's been following us around throughout our time on this earth."[11]

In the grand scheme of lice, you don't need to worry about ninety-nine percent of them. The ones you may interact with are also fairly easy to rid yourself (or your things) of. They have found their niche, be it with fungus or blood, and are rarely detrimental to the host. Not saying that everyone wins, necessarily, but it could be a lot worse. And those bark lice? Cute and completely harmless. To more bark lice!

QUICK FACTS: LICE

- Most lice like clean environments.
- The term *nitpicking* comes from the act of picking the small louse eggs (nits) from hair and how long that takes.
- Shaving your head will not get rid of head lice. You'll still need to pick nits and use shampoo recommended by doctors.
- Book lice are the original "book worms" and will eat mildew and mold that forms in moisture-rich books. Keep them books dry!
- While they may be annoying, parasitic lice are much easier to get rid of than some pests. Home remedies do not work, so I recommend going to the doctor to get the correct cream or shampoo.

11 Ari Daniel, "Lice DNA is a Revealing Textbook of Human History," NPR, November 8, 2023, https://www.npr.org/sections/goatsandsoda/2023/11/08/1211528184/having-lice-aint-nice-but-they-tell-our-story-concise-and-precise.

CHAPTER 3: CADDISFLIES (ORDER TRICHOPTERA)

*C*addisflies are another example of aquatic/semiaquatic invertebrates. The larvae and pupae live their life in water, while the adults are terrestrial. Most of the 14,500 identified species live in freshwater. There are a few species who live in brackish water or in the ocean.

Fossil records of caddisflies are sporadic. The earliest fossil is from 230 million years ago. Scientists even found a specimen in amber that has a parasitic wasp inside. Those types of fossils/encasements are very rare.

Caddisflies are in a sister group with butterflies and moths (order Lepidoptera). They split on the evolutionary tree recently (but by "recently" in evolutionary time, it means like 50 million years ago). The adults look fairly similar to a moth. They may even be difficult to tell apart when flying by you. They are also often attracted to artificial light, like moths.

Adult caddisflies are not great fliers and don't spend a lot of time very far from water. Many species as adults don't have mouths and live for a very short amount of time. For this reason, they lay their eggs close to or even in the water.

How Caddisflies Shape Our World

Caddisflies are important for quite a few reasons. One, they are invaluable as food for fish. Fish eat them at all stages of the bug's development, accounting for a large portion of the fish's diet. Since they are so tasty to fish, people have been using lures that resemble them when fishing for centuries. For example, in a 1653 book about fly fishing, the author Izaak Walton talks about using caddisflies as bait. And fish aren't the only animals that find them yummy. In Japan, a specific species of larva is considered a delicacy.

Caddisflies also are important bioindicators for water health. In fact, three semiaquatic insects help determine levels of pollution in waterways. This test is called EPT, which Land Air Water Aotearoa (LAWA), defines this way:

> EPT stands for Ephemeroptera (mayfly), Plecoptera (stonefly) and Trichoptera (caddisfly) which are macroinvertebrates that are sensitive to water pollution. Because these species are generally found in streams with good water quality, their abundance can give us an idea about how healthy a stream is.[12]

Scientists determine the EPT by counting the total number of these macroinvertebrates, divided by the number of species found, then multiply that by one hundred. The higher the percentage, the better water quality. Around the world, this metric determines the health of freshwater systems.

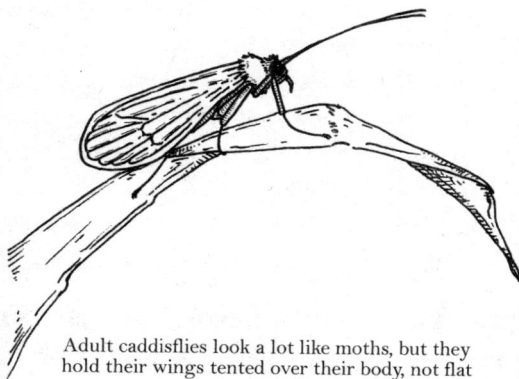

Adult caddisflies look a lot like moths, but they hold their wings tented over their body, not flat out to the side like moths.

The larvae of caddisflies are also fascinating. Many species create a little tubular house around them for protection, using silk they produce and pieces of whatever they find around them. Some use sand or small rocks, while others use leaves, algae, moss, shells, or pine needles. One of the saltwater species uses coral to create their traveling homes. The results are almost unreal-looking. They are so well put together and gorgeous. And that is just with what they can find in nature.

---

12 "EPT," LAWA (Land Air Water Aotearoa), n.d., https://www.lawa.org.nz/learn/glossary/e/ept.

Researchers have also been looking into some surprising aspects of caddisfly larvae and their interactions with microplastics.[13] In order to make their mobile homes, the larvae break down material to make it fit their structures. Because of this, they may break down plastics into microplastics which can have negative effects on water and human health.

There are also issues with caddisflies using plastic in their casings. First, plastics tend to be lighter than organic materials. This could potentially lead to buoyancy issues, causing them to float more than stay on the bottom rivers and creeks. That has the potential to make them more vulnerable. Second, plastics tend to be more conspicuous than leaves or other organic materials, making it easier for predators to spot them. Third, the predator that eats them is then ingesting pieces of plastic, which leads to issues over time.

Breaking down plastics is a topic that researchers have been looking into for quite some time, as we produce absurd amounts of plastic that then gets into animal bodies via water, food, and the environment. Some researchers are trying to find animals that can help in breaking down plastics, whereas others are trying to determine the overall effects on animals. A review of studies into microplastics states, "MP [microplastic] exposure can lead to physical harm, chemical exposure, inflammatory responses, and even behavioral changes in animals. Chronic exposure to MPs may have long-term health implications."[14]

How Humans Can Interact with Caddisflies

Shifts in habitat can be devastating to an entire species of caddisflies. In 2024, the Center for Biological Diversity petitioned to get the wonder caddisfly on the Endangered Species List. This species lives in a 110-

13 Katey Valentine, Richard Cross, Ruairidh Cox, Gina Woodmancy, and Alistair B. A. Boxall. "Caddisfly Larvae Are a Driver of Plastic Litter Breakdown and Microplastic Formation in Freshwater Environments, Environmental Toxicology and Chemistry 41, no. 12 (2022), https://doi.org/10.1002/etc.5496.

14 Eunju Jeong, Jin-Yong Lee, and Mostafa Redwan. "Animal Exposure to Microplastics and Health Effects: A Review," *Emerging Contaminants* 10, no. 4 (2024), https://doi.org/10.1016/j.emcon.2024.100369.

yard area of a river in Oregon, right under a waterfall. This is the only known location of these caddisflies—330 feet of space, that's it. It reminds me of the Whos who live on a dandelion. They are extremely vulnerable should anything happen to the river: development, pollutants, etc. We have the ability to protect these wonder caddisflies, so the Center is trying their damndest using the means available to them. Here's to hoping the wonder caddisfly continues to thrive in its micro-habitat.

When searching for caddisflies on Google, one of their suggested "Things to Know" is extermination. Let me be very clear: if these show up in your yard, they should not be exterminated. They are extremely important to waterways and are experiencing drops in population. Having caddisflies where you live is an amazingly good thing. The adults may swarm around lights but that is a nuisance at best. Seeing them should make you feel good about the water around you: it's healthy and it's large enough to sustain a population. Caddisflies are also food for fish and other animals. If you kill them, you're an asshole. Period.

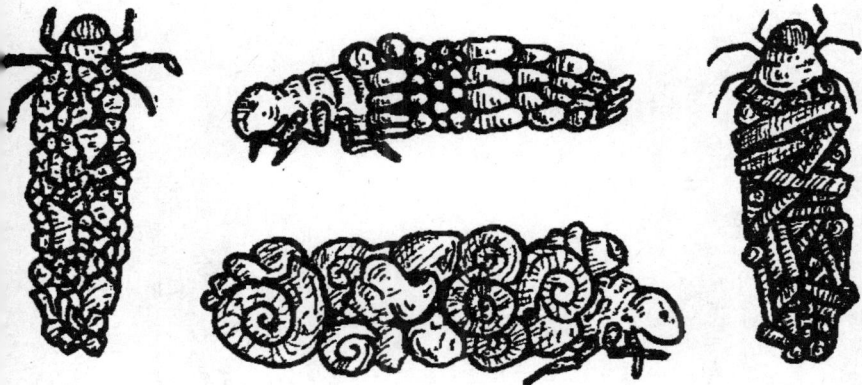

The caddisflies on the left, bottom middle, and right are all wild, using rocks, shells, and grasses. The one in the top middle is using precious materials like those that may have been given to them by the artist Hubert Duprat.

How Human Culture Is Shaped by Caddisflies

An artist named Hubert Duprat wanted to see what caddisfly larvae would do if they were only given precious gems and stones to make their tubular homes. The result was stunning tubes made of gold leaf, pearls, and other precious materials. These tubes sell for quite a lot of money, but the caddisflies don't see any of that cash. (Hope they live their best lives while in his care.)

In the time since then, people have either collected wild ones or had captive species whose delicately-created homes are then sold as jewelry on sites like Etsy. They can go for around $60 for a necklace or pair of earrings. If you want to watch a video of them putting their structure togethers, *This Is Colossal* has footage that shows Hubert Duprat's caddisflies at work.[15] Having them create these golden tubes helps create awareness around an often overlooked order. Caddisflies are integral to the ecosystem, but most people don't know what they are.

Quick Facts: Caddisflies

- Caddisflies are used as indicators for water health. Their presence is a good sign, and entomologists worry when they are not there or are in small numbers.
- The larvae create elaborate walking homes out of materials available to them that are stitched together with silk—plants, moss, rocks, snail shells, you name it.
- A species can be incredibly specific and live in a very small area. One vulnerable population is only found in a 330-foot stretch of river in Oregon.
- They are a sister group to lepidopterans (butterflies and moths).
- They are an extremely important food source for fish.

15 Christopher Jobson, "Artist Hubert Duprat Collaborates with Caddisfly Larvae as They Build Aquatic Cocoons from Gold and Pearls Colossal," Colossal, July 25, 2014, https://www.thisiscolossal.com/2014/07/hubert-duprat-caddisflies/.

CHAPTER 4: LACEWINGS, MANTIDFLIES, OWLFLIES, AND ANTLIONS (ORDER NEUROPTERA)

*T*he order Neuroptera (from the Greek for "sinew" and "wing") includes the lacewings, mantidflies, owlflies, and antlions—so named because of the intricate line work of all of their wings. The largest group is the lacewings, the one people see most often in urban spaces.

Neuropterans have been around for about 250 million years. Fossil records show an extinct species that was 9 inches wide. Not quite as big as extinct dragonflies or extinct millipedes, but still rather large. These days, most lacewings are about 1 inch long, but antlions can get to 5 inches wide (and kind of look like damselflies). Across all types of neuropterans, there are 6,000 known species.

HOW LACEWINGS, MANTIDFLIES, OWLFLIES, AND ANTLIONS SHAPE OUR WORLD

Almost all species have four wings of equal length, and have them tented over the body, similar to katydids. Adults may or may not have mouths. If they do eat as adults, they are rarely predatory and rely on pollen or nectar, whereas the larvae of almost every species is predatory.

Adults are very well camouflaged, often green or brownish in color. They go through diapause (similar to hibernation) during the cold months and often change colors with the leaves: they become brownish in the winter but become green again in the spring. Adults have a very delicate body and wings with giant, bulbous eyes.

Like other well-camouflaged, nocturnal insects, they can sense echolocation with the "ears" on their legs. They use this information to hunker down until a bat has passed.

They already live in many places but as the earth gets warmer, their habitats are growing northward. They are not bad to have in your garden. Similar to beetles, entomologists believe that they are "mess and soil" pollinators: perhaps not meaning to pollinate but doing it by accident. If you don't see the adults, you may see their eggs.

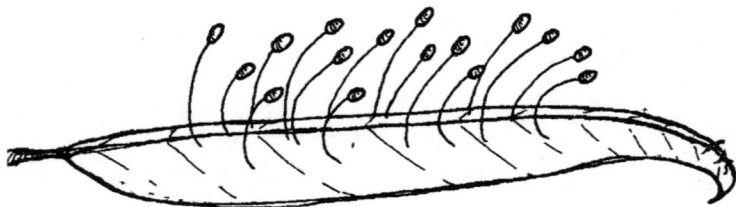

These eggs are so small, it is easy to miss them. Especially since they are often on the underside of leaves.

Lacewing eggs almost look like wisps of hair. The larvae are really the ones that are effective in your garden, at least as it relates to pests. They are serious predators who can eat an impressive amount of pests like aphids. Some lacewing larvae even carry around the bodies of the dead they've eaten on their back. Pretty hardcore. The larvae are also capable of spinning silk, which they use to make a pod where they feed, oftentimes on the underside of leaves.

They are also talented with their camouflage. The larvae use their little front leg grabbers to pull vegetation over them to hide and avoid predators.

Depending on the species, some larvae may carry around the exoskeletons of those they have killed on their backs. That's why they earned the name "trash bug." People didn't realize they were looking at a larval form of something and that the trash was actually the bodies of its foes. Badass.

Antlions are another neuropteran that you may have seen but not known what it was. The adults kind of look like a damselfly: long, thin bodies with long, thin wings. They can even be pretty large, upwards of six inches.

But it's not the adults that are most often recognized. Instead, what is most often seen is the burrow of an antlion larva. Some people call them "doodle bugs." Whatever you call them, they have distinct circular burrows most often seen in sand or loose dirt. Larvae burrow to form their pits. Small bugs, like ants, fall into these pits and have difficulty escaping because of the loose material of the walls. As they are struggling to get out, they continue to fall closer to the bottom. When the ant gets close enough, it is snatched by the antlion's mandibles (mouth pieces/hooks), dragged under, and sucked dry. Antlions are "sit and wait" predators because they hang out at the bottom of their burrows until something falls in. They don't run around looking for prey.

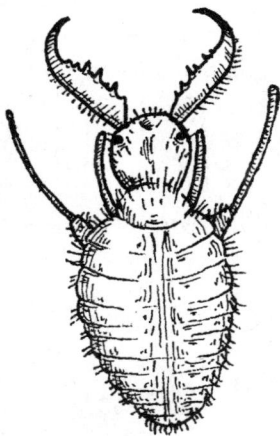

What an impressive set of jaws on a little chicken-nugget-looking body!

If the antlion larvae are in danger of being eaten, the larvae have been found to play dead, sometimes for up to an hour, in hopes of tricking the predator. Some insects, like parasitic wasp and fly species, will allow themselves to get caught so that they can lay eggs in the antlion.

Another neuropteran is the mantidfly. As adults, they share quite a resemblance to mantises but aren't related. The common name makes sense though. These insects show what biologists have termed "convergent evolution": two species who developed similar features but independently. Different times, different species. They both have raptorial front legs, which suits mantidfly adults, who are predatory.

As larvae, mantidflies live in spider egg sacs and feed off of them. They have some interesting behaviors. In order to eventually get to a spider sac, the larvae will wait for a spider to come by. When the spider does, they hitch a ride in anticipation of them laying an egg sac soon. If they hitch a ride on a male, they will have to wait until that spider encounters a female and transfer over. Since they must wait, in order to get food and continue to grow, they drink the blood of the spider. Not enough to kill the spider but enough to sustain themselves. When they do find themselves on a female and that female has finally made an egg sac, they move into that. They have also been found as parasitoids of bees, wasps, and scarab beetles. Sometimes they are food for others, including humans. You won't find them in markets as frequently as other bugs, but some people view the larvae as a delicacy.

The last insect in this order is called the owlfly. They got their name for their nighttime hunting, like owls (animals who hunt or are active at twilight are called *crepuscular*). The adults look like what you might get if a dragonfly and a butterfly had a baby. The larva looks like their antlion relatives and has similar behaviors. Compared to the others in the group, owlflies are strong and agile fliers, having some maneuverability that is similar to dragonflies. They are predators that feed on flying insects, whom they often capture mid-flight. Mating also occurs in flight.

I have never seen one of these in my life, but they are gorgeous. Or maybe I have and didn't know it wasn't a dragonfly.

While many of the larvae look the same, the adults are all fairly different and often resemble an insect from a different order. Adults: antlion, lacewing, owlfly, and mantidfly.

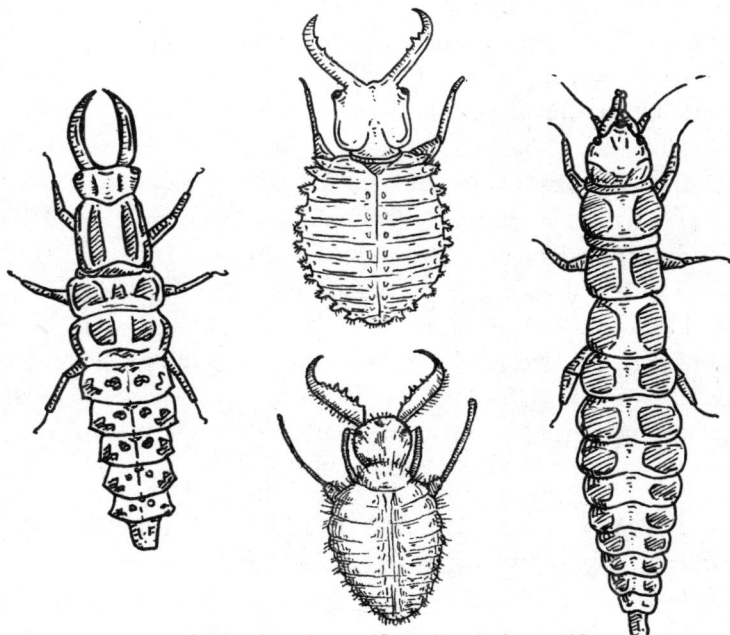

Larvae: lacewing, owlfly, antlion, and mantidfly.

How Humans Can Interact with Lacewings, Mantidflies, Owlflies, and Antlions

Lacewing larvae are great predators and are known to be helpful in the garden. They love eating aphids (alongside ladybug larvae), but they are often generalists and will eat whatever is available.

Some farmers and scientists have released them into non-native spaces in the hopes of them being natural pest control agents. That always has the potential to end badly, be it for the farmers, the public, the animals, or the plants. The North American green lacewing is native to the contiguous US, but not Hawaii. These lacewings have found themselves there (likely by accident—maybe on purpose in green houses) and biologists are concerned about their effect on local lacewing populations. If they become established in the wild, they could easily outcompete the native populations since they have fewer natural predators.

Humans may unknowingly see new or rare species all of the time. In 2012, an entomology PhD student at the University of Arkansas found what they thought was an antlion adult in the parking lot of a Walmart. The specimen was labeled as an antlion. In 2020, the student was teaching a class on Zoom and used the pinned specimen to describe attributes of antlions. They realized that it kind of looked more like a lacewing, and thus, his discovery was made in a live Zoom class. It was a specimen that had not been seen in the US since the 1950s.[16] (NOTE: It takes a very long time to get through specimens at a museum. I've pinned fifteen-year-old beetles at the museum I volunteer at. It is very common since they rely a lot on volunteers for the process.) Two antlion species were discovered in 2023, as was a new species of owlfly. iNaturalist and citizen scientists are helping find new species almost every year.

How Human Culture Is Shaped by Lacewings, Mantidflies, Owlflies, and Antlions

Because of the iconicness of their burrows, antlion larvae in particular (not the adult forms) have been found in popular culture. Some people claim that the Sarlacc pit monster in *Star Wars* is a reference to a doodle bug trap, but George Lucas has never officially stated that.

Antlions do appear in video games, often portrayed as being big enough to eat humans. Their larval form is surprising and has been described as sinister. There even is a Pokemon evolution based off of them (larvae and adult) as well: Trapinch, Vibrava, and Flygon. I don't necessarily see the resemblance but Trapinch (great name) does have behaviors similar to an antlion larvae.

In 2011, scientists from the Berlin Museum of Natural History found a "unicorn fossil." It is dubbed as such because the chances of finding one are about as great as finding a real unicorn. These fossils not only include the animal, but also capture the behavior of that (or

16 Adrienne Berard, "Rare Insect Found at Arkansas Walmart Sets Historic Record, Prompts Mystery," Penn State, February 27, 2023, https://www.psu.edu/news/research/story/rare-insect-found-at-arkansas-walmart-sets-historic-record-prompts-mystery/.

numerous) animals. The latter was the case for the amber found in 2011. Michael Ohl and his team found a spider. And there, latched onto the spider's underside just as its modern relatives do, is a mantidfly larva. They have had the same behavior, as it is seen now, for the last forty-four million years.[17] Obviously, their system is working well for them.

QUICK FACTS: NEUROPTERA

- Larvae of all neuropterans are vicious predators, but not to humans. You don't want to be an ant that crosses their path.
- Mantidflies look like preying mantises but are not related. This is an example of convergent evolution: two animals that have similar features but evolved separate from one another.
- Adult owlflies and antlions can easily be mistaken for dragonflies. They have equal-length wings and thin bodies.
- The Sarlacc pit monster was possibly inspired by antlion dens.
- They are all good for your garden, especially as larvae, and are then an important food source for larger insects or vertebrates when they are adults.

17 Ed Young, "Spider-Boarding Insect Preserved in Amber," *National Geographic*, March 28, 2011, https://www.nationalgeographic.com/science/article/spider-boarding-insect-preserved-in-amber.

CHAPTER 5: WALKING STICKS AND LEAF BUGS (ORDER PHASMATODEA)

One of the best moments in *A Bug's Life* is when they are trying to find Slim, the stick insect, and he yells out to them "I'm the only stick with eyeballs!" Thankfully, Slim had large, obvious eyeballs. In reality, the eyes of walking sticks are extremely difficult to notice. When I was a kid and we would go camping, without fail at least one would appear on my chair, doing it's little back and forth dance. The ones I saw as a kid were fairly small and delicate, but always adorable.

As of 2009, there are 3,000 known species of Phasmatodea (from the Greek for "phantom"). Species diversity is richest in tropical and subtropical regions. Many of the known species are native to Asia. They are most active at night (but even then, they move pretty slow). Females can lay upwards of 1,000 eggs at a time, and the young look just like tiny versions of the adults, which is a form of incomplete metamorphosis.

The longest insect in the world is a walking stick, reaching lengths of twenty-five inches (*Phryganistria chinensis*, found in China in 2014). They are fairly docile and do not harm humans (or anyone really). They have been kept as pets in China since the Han dynasty and are also found in US zoos

Giant Malaysian leaf insect, *Pulchriphyllium giganteum*

in the present day. They are a slow-moving family that eat vegetation, like flowering plants, almost exclusively. The largest species in North America is the Giant walking stick, who gets to be about seven inches long. That is large for North America but doesn't quite compare the ones in China. Leaf bugs, while related to stick insects, cannot be found in all parts of the world. They only live in Asia, Australia, and other small isles.

How Walking Sticks and Leaf Bugs Shape Our World

They are masters of mimicry. Leaf bugs have veins you would normally see in a leaf. Some species even go so far as to have fake bite marks to look like they are a leaf that has been munched on.

Stick bugs, or walking sticks, will have the same coloring as the trees they are on and may even have moss-like patches to further blend in with their surroundings. You may never see one simply because they are so well hidden. Leaf insects are flat compared to their other leaf relatives, katydids. Katydids have vertically-laid wings and most leaf insects have reduced wings (if any at all).

Mating is an interesting behavior in leaf/stick bugs because there often is no need for a male. The females can reproduce through parthenogenesis, meaning the female can lay an egg with no sperm and the females that hatch are the exact replica of their mother. Some species have only ever been found to have females. Some species do a mix of parthenogenesis and mating, and a few rely on mating between males and females. Often, males of the species are significantly smaller, an example of sexual dimorphism). In some cases, the female is fine with the male staying on her back for quite some time (upwards of seventy-nine days, which is very long and uncommon) because she has a higher survival rate when he is there. He's a stick bug shield and he is more likely to get eaten or injured than the female while they are in the mating position.

Researchers have witnessed competitiveness between males in some phasmatodean species. They describe the fighting as a type of boxing match. The males swing their topmost legs at each other while standing on their hind legs. A slow boxing match between two very tall individuals. It sounds like a hilarious sight.

Stick bugs never throw fists as a defense against any would-be predators. As mentioned above, they are very well camouflaged but have some additional ways of protecting themselves. Some have bright colors on their small wings to surprise a predator, hopefully long enough to make an escape. They have also been seen feigning death or falling to avoid being eaten. Some have an irritating spray they can shoot from their bodies at a predator. Similar to mantises, they also move in a swaying motion to imitate the movement of leaves to disorient predators. Mimicry is seen in a species whose young look very similar to ants. Ants are a big predator of small stick bugs, so they evolved to look like ants in hopes of tricking them (it works).

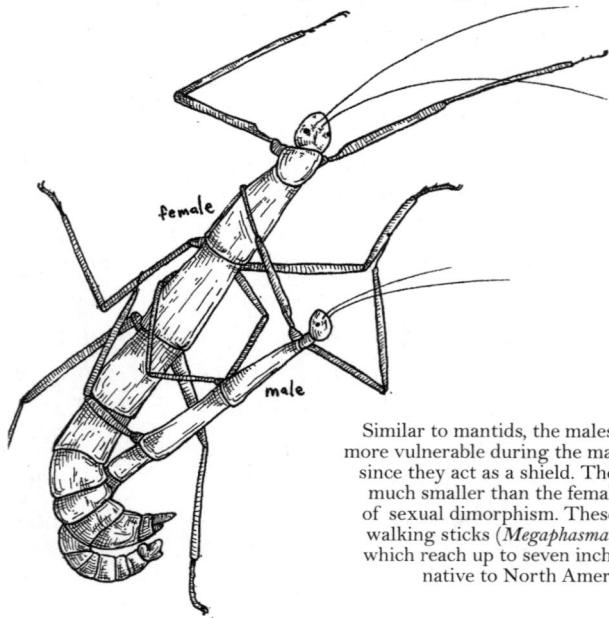

female

male

Similar to mantids, the males are much more vulnerable during the mating process since they act as a shield. The males are much smaller than the females, a type of sexual dimorphism. These are giant walking sticks (*Megaphasma dentricus*), which reach up to seven inches and are native to North America.

How Humans Can Interact with Walking Sticks and Leaf Bugs

There is much less known about leaf bugs in comparison to stick bugs. Scientists didn't discover the first leaf insect fossil until 2007.[18] It was found by a group of German researchers who dated the fossil as being 47 million years old. Much of what was found in the fossil is similar to what these creatures look like today. Another group of German scientists (with an American in the mix) found seven new species of leaf insects, all in 2023.[19] Since that group has started looking for species in tropical areas, they have identified around twenty new species in total. Undoubtedly there are still new species to identify. Or even some known ones that need to be reassigned.

In 2020, a Montreal researcher, Stéphane Le Tirant, got a clutch of eggs from a Papua New Guinea leaf bug. He knew that this was a species that had only identified females previously. He was expecting a few tiny female leaves; what he got was a single leaf and two unknown sticks who they discovered were males. It turns out that the males had been seen before but were given a different scientific name and were placed in a different genus. That species had only ever had males. Now, after one hundred years, they can rename the males and put them together as one species—females as leaves, males as sticks.[20]

Regardless of whether you have leaf bugs or stick bugs where you live, many rarely see one. They are masters of camouflage and are difficult to find. You may have passed hundreds in your lifetime and had no idea. If you do see one, be gentle, as they are delicate, friendly, and slow-moving.

18 Lisa Zyga, "Scientists Discover First Fossil of a Leaf Insect," Phys.org, February 8, 2007, https://phys.org/news/2007-02-scientists-fossil-leaf-insect.html.

19 University of Göttingen, "Curious and Cryptic: New leaf Insects Discovered," Phys.org, August 28, 2023, https://phys.org/news/2023-08-curious-cryptic-leaf-insects.html.

20 Sabrina Imbler, "He Was a Stick, She Was a Leaf; Together They Made History," The New York Times, December 1, 2020, https://www.nytimes.com/2020/12/01/science/leaf-stick-insects-phyllium-asekiense.html.

While leaf bugs are not native to the US, their inclusion in zoos as well as in the pet trade are common. Unlike other bug species kept as pets, as of yet, no researchers have found a correlation between exotic leaf bug pets and species decline. They do reproduce rapidly and since the females don't need a male, their numbers can quickly get out of hand. In Asia, they are often kept because they are believed to represent fortune and good luck (like crickets).

How Human Culture Is Shaped by Walking Sticks and Leaf Bugs

Research has also revealed another amazing thing about stick and leaf bugs: If a female stick insect is eaten, her eggs may survive through digestion and still hatch. A group of Japanese researchers published their findings in the journal *Ecology*. They fed the stick eggs to birds and after being passed, a third of the eggs still hatched. Their eggs are covered in the same substance that makes kidney stones in humans, so the exterior is very hard.[21]

Another group of researchers are using stick insects to create robots that can move in similar ways. At Bielefeld University, they attach motion capture tags to the front legs of adult stick bugs. They then record the movements to determine how to best create joints in robots. Insects have more leg sections and joints than we do, so determining how the insects use them can assist them in making robots that can walk smoother. Insects move very efficiently, and scientists are hoping to make robots using the knowledge gained from studying their movements.[22]

21 Jillian Mock, "Neat! Stick Insect Eggs Can Survive Being Eaten by Birds," *Audubon Society*, June 5, 2018, https://www.audubon.org/news/neat-stick-insect-eggs-can-survive-being-eaten-birds.

22 James Gorman, "Stick Insect Helps Scientists Study How Animals Move," *The New York Times*, February 15, 2016, https://www.nytimes.com/2016/02/15/science/stick-insect-helps-scientists-study-how-animals-move.html.

QUICK FACTS: WALKING STICKS AND LEAF BUGS

- They are basically just the bug equivalent of cows: they hang out and eat vegetation all day, posing zero threat to humans.
- Most females can reproduce without needing to mate with a male. They give birth to exclusively female young in a process called parthenogenesis.
- Similar to mantises, stick and leaf bugs move side to side in an attempt to mimic the natural movement of vegetation in the wind.
- People have been keeping stick insects and leaf bugs as pets since the Han dynasty (which started in 206 BCE).
- Their eggs are covered in the same material that kidney stones are made from and can survive a trip through the digestive tract of birds.

Chapter 6: Thrips (Order Thysanoptera)

*T*hrips are often-unseen tiny little insects. Though they are tiny, they are important little dudes.

Thrips comes from the Greek word for "wood worm," though these creatures have also been called corn lice, corn fleas, and corn flies.[23] They are none of those, as they are insects in their own order. The name *thrips* refers to a single thrips or several thrips, similar to deer and moose. That s is always there.

They are very tiny, normally around one millimeter. They have big mouthparts (mandibles) for their size, with a sucking mouthpart underneath. The mandibles are used to cut open their prey to create a hole

Often go unnoticed since they are so tiny

for their sucking mouthpart. They stay tiny their entire lives, even though they continue to molt while in adulthood (very uncommon for insects, more common in arachnids).

There are around 7,700 species worldwide. Most are fungivores (they eat only fungus) or detritivores (they only eat decomposing material). Some however can be considered pests as they do eat some crops. Others may even be carnivorous: a few species have been found eating the larvae of wasps, mites, and moths.

Some thrips males are known to be very defensive of their eggs, an uncommon practice with invertebrates. Males fight for the attention of females (which is very common), but sometimes a smaller

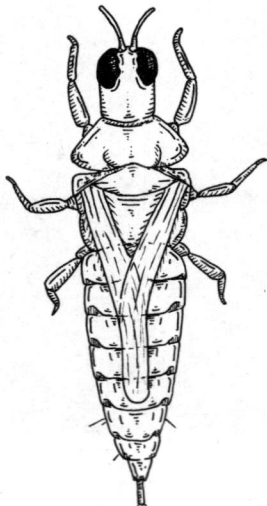

23 Peter Marren and Richard Mabey, *Bugs Britannica* (Chatto & Windus, 2010).

male will sneak in to mate with the female while the other males are fighting. They don't always need to mate as females are capable of parthenogenesis, similar to stick insects.

How Thrips Shape Our World

Almost everything you can find online about thrips is negative— articles with titles like "How to Get Rid of Thrips for Good," "How to *Actually* Get Rid of Thrips," and "Thrips: How to Identify and Deal with Them."

Of all the species, about ten total are seen as pests, which is a very small percentage of their numbers. They can transmit diseases from plant to plant, as well as create scars that may be unattractive to buyers or farmers. Many species have spread to other parts of the world through trade routes, as they are tiny and often on produce. These pest species are a big problem once they move into a new space because they do not have any natural predators. Add in the ability to be able to produce asexually, and their numbers can quickly get out of hand because they are left unchecked. Sometimes they don't need the help of humans to spread. Many species use wind to carry them to new places. An Australian thrips species gets carried over 1,000 miles to New Zealand when they have their migration every year.

Most of what thrips do is positive. They are important pollinators, especially for certain types of plants: heather plants in the Northern Hemisphere, a genus of economically important trees in Malaysia, the Panama rubber tree, several endemic species in Australia, and some tropical plants. Some research supports the idea that thrips are particularly fond of yellow and white flowered plants. They carry pieces of pollen on their backs. The number of pollen pieces is much smaller than what a bumblebee can carry, but thrips do still move pollen from flower to flower. Fossilized amber was found by researchers in 2012 that showed a thrip carrying pollen from 110 million years ago, showing they have been pollinating for quite some time.[24]

24 Enrique Peñalver, Conrad C. Labandeira, Eduardo Barrón, and Carmen Soriano, "Thrips Pollination of Mesozoic Gymnosperms," Proceedings of the National Academy of Sciences

A tiny, unsung pollinator

How Humans Can Interact with Thrips

While thrips may be small, they can still bite humans. It's not a medically significant bite, more like a pinch. If you didn't know they were there before, you will after they bite. And they are so tiny, you will likely kill them in the process of discovering them. Also, I'm sure it could go without saying, but they do not want to be on you, nor do they purposefully go after humans.

Scientists at the USDA are hoping that they can help against the spread of the Brazilian peppertree in Florida. The tree is not native to Florida and is spreading quickly. Imported thrips (who exclusively eat the tree) are being given out by the USDA to anyone who might be interested in releasing them on their property.[25]

At this stage, the end result is unknown. Humans are known to release an invertebrate to take care of a pest issue only to discover that while it may help curb one thing, it causes issues with something else. That isn't to say this will be the case against peppertrees, but it is often a solution that isn't thought through in regards to long-term effects.

Some species are known to create galls to live in, some small and some very large. The sizes of the gall house can range from containing a few thrips to large, complicated structures that hold up to 10,000 adults. In social species, the first round of eggs that hatch is turned into guards. These guards help with construction of the home, as well as keep on the lookout for predators who may be coming their way.

Oftentimes, thrips (even in large numbers) are not harmful to gardens. If you do want to avoid getting them or are running into an issue

109, no. 22 (2012), https://doi.org/10.1073/pnas.1120499109.

25 US Department of Agriculture, "Thrips Show Promise in Controlling the Invasive Brazilian Peppertree in Florida," October 12, 2022, https://www.ars.usda.gov/southeast-area/fort-lauderdale-fl/iprl/docs/information-about-schinus-terebinthifolius-brazilian-peppertree-and-recent-biocontrol-release/.

with them, current research states that the best way to control them is by using red lights or red nets to catch them. Thrips adapt very quickly to pesticides, so those are not recommended since the pesticide would no longer kill the thrips but might kill other beneficial bugs. Also, scientists do not recommend the use of pesticides in general. Even if it does kill a so-called pest in your garden, pesticides are dangerous to humans, pets, ground water, and many other animals. Like microplastics, continued exposure to pesticides is harmful for everyone. Think of that before you try to use bug killer for thrips on your plants.

How Human Culture Is Shaped by Thrips

Because they are so small, thrips have seemed to escape widespread notice or inclusion in stories, folklore, and art.

Jordan Pérez, a poet, includes them in one of their poems called "Knockout Rose." I highly recommend it, not only because it's beautiful but because it shines some love and recognition onto a less-discussed insect.

Quick Facts: Thrips

- They are very tiny (one millimeter) and stay that size their entire lives
- They like white and yellow flowers. After they visit a flower, they carry pollen on their back to other flowers.
- While they are pollinators, thrips do not carry nearly as much as other pollinators because they are so small. They do their part though!
- They are fighting an invasive tree in Florida. People can ask the USDA for thrips to put on any of the invasive trees on their property
- They adapt quickly to pesticide, so if you want to get rid of them in your garden, try to use non-chemical methods like red lights (shining on plants) or red nets (placed over plants) to deter them.

CHAPTER 7: SILVERFISH (ORDER ZYGENTOMA)

*B*ehold these glorious, ancient creatures—who are often found in your bathroom. The first time I recall seeing a zygentoman, it was on the wall by the toilet paper when I went to reach for it. Definitely gave me a shock.

Order Zygentoma includes one of the oldest insects: the mighty, four hundred million-year-old silverfish, evolved before insects developed wings. These shiny silver creatures are often blind (or have really bad eyesight) and prefer the dark. They have no wings but can run very fast. They have three distinctive tail bristles that stick out the back with long, thin antennae out the front. They love a steamy shower as much as the next guy (hence the bathroom appearance—though they'll appear in really any humid place).

Silverfish have adapted very well to living in human dwellings, eating what we leave behind or what the humidity helps grow. Also, they have found a way to survive without needing wings. Their only defense mechanism is speed. Their name comes from their shimmery, plated covering. The silver segments almost look like fish scales. They also have an interestingly shaped body that resembles a fish: broader at the head and tapered towards the back.

Firebrats have very similar body structures and lifestyles to silverfish. They appear all over the world but can be found in less humid places. When in homes, they are most often found near heat sources (hence the name). Their appearance differs in that they are brown and more likely to blend in with leaf litter or natural environments.

There are a few species in this order who are troglobites, meaning they live underground, most often in caves. They may be specific to one cave and belong to their own sub-species. With that amount of specificity, some species are easily at risk for extinction should something happen to their environment.

The three tails sticking out of the back of all silverfish are called cerci, found on many bugs. For some, cerci serve various functions. For this order, they are to help with feeling what's around them. To be clear, though, cerci are not stingers and will not harm you.

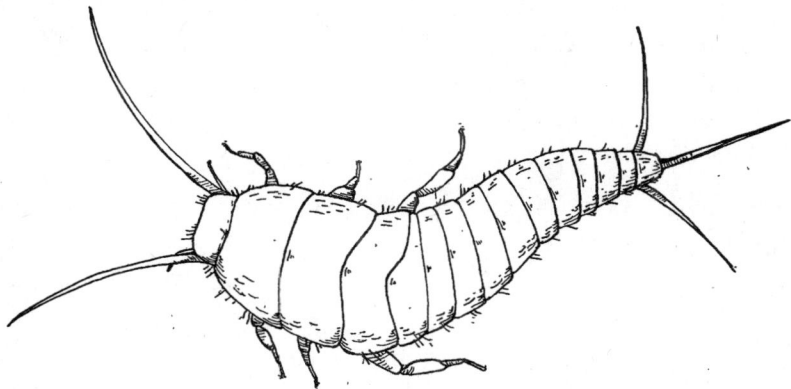

Silverfish (*Lepisma saccharinum*) can reach up to one inch in length. The three tails out the back are not stingers and are harmless.

How Silverfish Shape Our World

Of the thirteen species in the US, only three silverfish are thought to be native. Many non-native species are now worldwide because of human movement. An estimated seventy percent of silverfish in all locations are non-native. Surviving well in both their native and non-native habitats, they've also adapted to need very little food and water. If there is access to water, they can live a whole year without eating. What they do eat is often fungus or detritus, so they eat the stuff that humans don't want to see in their houses.

Their lives can also be surprisingly long for an insect. Some species can live up to eight years. Some have also adapted to living with other animals, like termites and ants. They may live in their nests with them in an example of nest commensalism. They don't produce large numbers of young. A female only lays about one hundred eggs in her lifetime, so if she lives for eight years, that is only about twelve eggs each year. And their mating dance is pretty adorable. It is a four-step process.

1. The male and female start face to face, vibrating antennae on each other.
2. Then, the female chases the male around.
3. Then, they stand side by side while the male vibrates his tail against the female before she takes his sperm.
4. He places his sperm packet on the ground, and the female then puts the packet in her reproductive organ

How Humans Can Interact with Silverfish

They love to eat what we leave behind, and they prefer to do it in sauna-level humidity.

Even with this adorable dance and lack of harm to humans, people are not often excited to see silverfish in their houses. Good news is, they aren't a disease vector and infestations are rare. In general, silverfish are not bugs you need to worry about. Even if you see one, chances are you don't have an infestation. If you'd prefer to not see them, anecdotally people say they don't like cedar oil. Placing that along floorboards may make them less likely to enter your home. In very dry environments, they are less likely to be a problem, apart from bathrooms or other unnaturally humid parts of the house. They thrive in humidity that is over seventy-five percent. In the end, the takeaway, hopefully, is that they may look creepy but are harmless.

Another insect in houses that makes some people anxious is the house centipede (which will be talked about in more detail later in this book); however, they love eating silverfish and other bugs that are often found in human dwellings. So if you see one, they may be to thank for low silverfish numbers in your home.

Researchers are currently studying this order to see if they can make any advancements in biofuel technology. Similar to termites, they can break down cellulose with their bioreactor of a gut, which is of great interest to humans trying to create commercial biofuel production procedures. As of yet, scientists have not figured out how the order can be used but continue to study the process in the hopes of making breakthroughs.

How Human Culture Is Shaped by Silverfish

The fact that silverfish don't show up much in culture is crazy, considering how long they've been around and how often they are associated with human dwellings. They are nocturnal and aren't a fan of lights, so perhaps they are not often seen because of that.

Jim Lichtenberg has an interesting poem about seeing a silverfish stuck in his bathtub. In the end, he smashes it to avoid taking a bath with a bug. Fair sentiment, but it's also easy (and not dangerous) to place silverfish outside the tub and allow it to crawl back under the floorboards.

In the video game *Minecraft*, silverfish will randomly appear and sometimes attract other silverfish, eventually turning into a "swarm." They don't really swarm, but there can be large numbers of them in one space if conditions are ideal. They are considered pests even in video games.

They were also an early "book worm," like book lice, because they like cellulose and have thus been known to eat pages. Some artists use the pages eaten by them and other "book worms" as art pieces or compliments to pieces they are creating. The pages have winding small tunnels created by the silverfish and other cellulose-eating beings that are not replicable by the human hand. Pretty cool looking, if not frustrating or completely devastating to the book owner.

QUICK FACTS: SILVERFISH

- Silverfish are the oldest recorded insect and haven't changed much in physical appearance since they showed up four hundred million years ago.
- Silverfish eat cellulose, as well as dandruff and mold. They themselves are not predators. They are eaten a lot by other predators, like earwigs, house centipedes, and spiders.
- Silverfish, despite the name, are not able to swim. They do often get stuck in bathtubs. (Check out Jim Lichtenberg's poem about that experience.)
- Silverfish almost never bite humans. There are a few random cases, but it is so rare that people say they don't bite.
- Silverfish are not harmful to you or pets. Your animal won't be poisoned if they eat one, but they may get an upset stomach if they eat too many.

CHAPTER 8: EARWIGS (ORDER DERMAPTERA)

L et's address the most egregiously false myth about earwigs first: They do not crawl inside your ear and make you go crazy. These 208 million-year-old, strange-looking bugs have no interest in your orifices. Their common name, earwig, may actually be a corruption of "earwing," since their uniquely shaped hind wing resembles a human ear. Their cerci also look like the instruments once used to pierce ears. The legend of them crawling into ears has been common since Anglo-Saxon times, so shaking that myth will be difficult.

Colorado was the first place I ever saw one and I must admit, I found the cerci to be intimidating. But then I remembered that I'm a giant to him and it was cold outside so I let him stay. [26]

The most commonly sighted species in North America is a non-native species, called the European earwig. Native species like the ring-legged earwig can be found but are not as common as the European species.

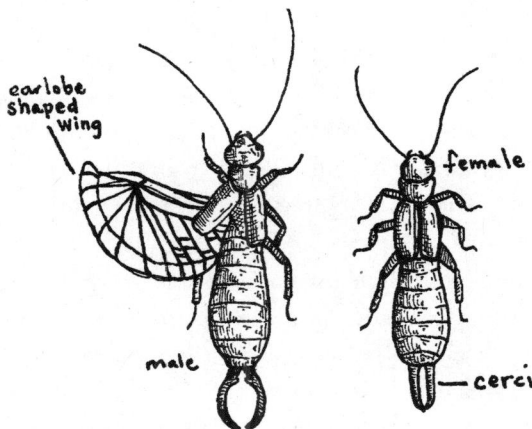

Males and females look fairly similar, but males have larger, more curved cerci.

26 I say him because of the shape of his cerci.

How Earwigs Shape Our World

Earwig wings are unique: semicircular, a shape not found in many other species. They fold fanlike beneath the front pair when the insects are at rest. The wings are also pretty short, so earwigs are capable of flying but are not very good at it. They prefer to just run away. If they need to defend themselves in some other way, some species have scent glands that squirt a foul-smelling yellowish brown fluid about four inches (impressive since they are normally an inch or shorter).

Similar to stick insects, when born, earwigs look like smaller versions of their parents. This is a type of incomplete metamorphosis, where they have multiple instars before adulthood.

Mothers are very caring to their young until they've molted a few times. Mothers will groom and defend them, as well as feed them regurgitated food. If they don't leave quickly enough after that stage, however, she may turn on them (i.e., eat them).

Most of the time, earwigs are omnivorous, eating detritus, fruits, and insects. They eat the silverfish discussed in the last chapter.

How Humans Can Interact with Earwigs

Similar to silverfish, seeing one does not indicate you have an infestation. They rarely infest a place and are often loners, passing through your place. They also do not carry any pathogens, so they aren't dangerous. They are also less likely to enter your home if you don't have rotting wood or food inside, or if your other bug populations are under control. The winter is an exception as they may travel inside to escape the weather.

They also cannot bite humans, but may pinch with their cerci. That hasn't stopped other countries from giving them exaggerated names related to their pincers. In Japanese, their common name translates to "penis cutter," which may have something to do with them often being found in bathrooms and their cerci looking scissor-like. In Spanish, they are called "little scissors." Again, just to clarify, their cerci are not sharp enough to pierce your skin. Their cerci are used for many other things, like capturing prey, defending themselves, fighting (between males at least), aiding in copulation, and helping them fold their wings.

It is possible to have a large number of earwigs present in your house that can cause many surprises for tenants. A friend of mine had so many in their home that they once went to brush their teeth and realized an earwig was on the handle of their toothbrush. That was the last straw for them. They didn't know what else to do, so they called an exterminator.

I get it—that is totally understandable when you feel like you can't seem to go anywhere in your home without seeing them and they show up in unexpected places. It is especially hard if you are renting as the upkeep falls on the landlord instead of the actual person living there.

In cases like these, I would not recommend spraying. Instead, try these steps to reduce the number of insects you see, limit the amount of chemicals that can harm you and pets, and keep the insects that do live in your home in smaller numbers happy and healthy. Even "safe" pesticides may not actually be safe for us or the planet. Earwigs are in your home because conditions are perfect for them. It isn't their fault—it's ours and there are things we can do to make sure they stay out of our space.

- Remove leaf litter and anything that holds moisture from window wells. This will also help reduce numbers of other bugs like silverfish and millipedes.
- Be sure to check for leaks or dripping water.
- Be sure that all floorboards and cracks are sealed.
- Remember to clean bathrooms as much as possible to reduce the amount of food created by humidity and things like human/pet hair (that equals a yummy snack for earwigs).

Outside the house, earwigs are sometimes known to be pests, particularly of soft-fleshed fruits like peaches. While they are pests themselves, they can sometimes help with other pest species. Their behaviors kind of cancel each other out.

Earwigs are also another insect that has been transported around the world because of human travel. Some of these species are beneficial. In Washington, researchers thought that they may be useful on apple farms since they are known to eat woolly aphids. Earwigs can't bite through the apple peel so they eat the aphids instead. Washington State University entomologists collected thousands of them from peach orchards and gave them to apple orchards where they educated farmers on why they were so useful. Farmers can reach out to be part of the study.[27]

In small household gardens, earwigs are likely to be found in soil, as well as potentially eating plants. They cause minimal damage, so there is no need to fret over their presence.

In rare cases, they can cause great damage. Again, that is highly unlikely. You may sometimes find them sleeping in plants or flowers, similar to some species of bees. Their lil' butts poke out of the flower. It's extremely adorable.

Lil' butt. Boop!

How Human Culture Is Shaped by Earwigs

In culture, earwigs show up because of the lore that surrounds them. James Joyce had a thing about earwigs and mentioned them in *Finnegan's Wake* and *Ulysses*. They have appeared in children's books (Roald Dahl), poems (including a Tennyson) and TV (a *Night Gallery* episode written by Rod Sterling). Sometimes, they are portrayed as

27 Eliza Billingham, "Perfecting Eden's Fruit Takes Geneticists, AI, and Some Earwigs," *Inlander*, October 11, 2023, https://www.inlander.com/food/perfecting-edens-fruit-takes-geneticists-ai-and-some-earwigs-26790559.

common animals seen in homes or on plants, and sometimes they are portrayed as dangerous creatures (which they aren't).

Part of their legend also says they burrow into your brain and lay eggs there. No validity to the myth at all. They are just weird-looking, little scavengers who are much more scared of you than you are of them.

Scientists are fascinated by earwigs' wings. When a group of researchers tried to recreate a paper model of the wings, it was not possible. Their wings fold in ways that don't compute to paper. They basically have springs that allow their wings to pop out quickly. Researchers are trying to determine how they could use the wing design to create useful products for humans: compact electronics, pop-up tents, and portable solar panels.[28]

Quick Facts: Earwigs

- They do not crawl inside your ear, nor do they burrow into your brain. Their name comes from the shape of their wings, which look like earlobes. Their name was originally "earwing."
- The pincers on their butts may pinch humans but only in situations where earwigs feel threatened). The pincers are really just for smoothing out their wings, aiding in sex, fighting between males, and making themselves look big.
- Seeing one does not indicate an infestation. If you see lots of them, you may have a leak somewhere, as they love moist wood and spaces.
- They are pretty good hunters who eat other insects that may be considered unwanted like aphids and silverfish. They are also important food for house centipedes and spiders.
- James Joyce has a thing for earwigs and mentions them in numerous works.

28 Douglas Main, "How Earwigs' Wings Break the Laws of Origami," *National Geographic*, November 13, 2018, https://www.nationalgeographic.com/animals/article/earwig-origami-wings-how-they-work-insect-flight.

Chapter 9: Scorpionflies and Hangerflies (Order Mecoptera)

*W*hile given the name *fly*, these insects are not part of the order Diptera (true flies). They are in an order all their own called Mecoptera, meaning "long wing"—known for their uniquely long faces and including more than six hundred species.

Both larvae and adults scavenge on dead bugs, but adults may eat pollen as well. In mating rituals, the male often brings the female a dead bug as a nuptial gift. If the female likes the gift, they mate.

Males will fight over females. If the male loses, he will take a different approach to finding a mate. So, instead of a nuptial gift, he may try to mate with a female while she is eating (and distracted) or try to force himself on her. Those attempts may not have as high of success as the nuptial gift, but male scorpionflies still try.

Male scorpionflies have a bulb at the end of their tales—deceivingly similar to a scorpion's tail (hence their common name); however, they do not sting and are incapable of doing so. Some scientists believe this bulb may be an example of mimicry. That is, when a species evolves to have similar "warning" markings of an animal that is venomous or poisonous. It's all for show but may make predators less likely to attack if they've encountered a scorpion before. Entomologists believe that butterflies, flies, and caddisflies evolved from the order Mecoptera, as they are further back on the evolutionary tree. The earliest fossil of scorpionflies is from about 280 million years ago.

How Scorpionflies and Hangerflies Shape Our World and We Can Interact with Them

Most species live in tropical or temperate places, but there are such things as snow scorpionflies. A new snow scorpionfly was discovered in Alaska in 2013 via a Facebook bug group, and another new species was discovered in Nepal in 2022.

The Nepalese species has an appearance that is quite different from most in the order. The newly-found species has an extremely elongated abdomen, which gets very thin and then has pincers on the stinger-looking organ at the end of the abdomen. The stinger is still unable to do any stinging but does look more intimidating.

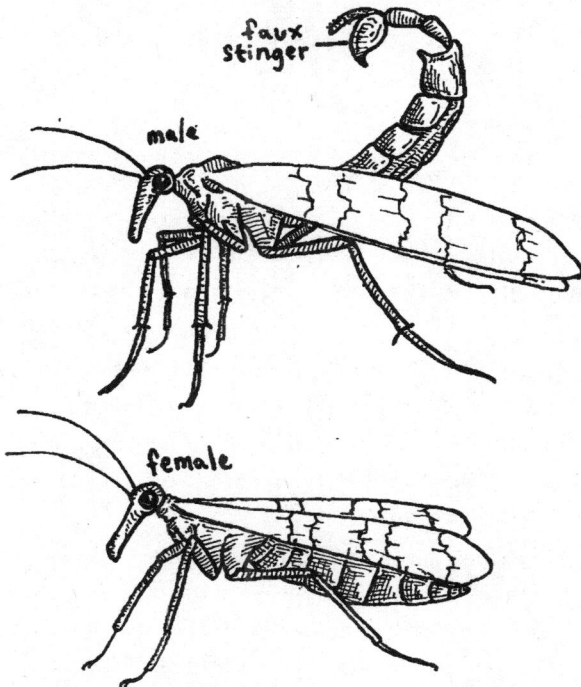

Notice the difference between the female abdomen (no stinger-looking appendage) and male abdomen.

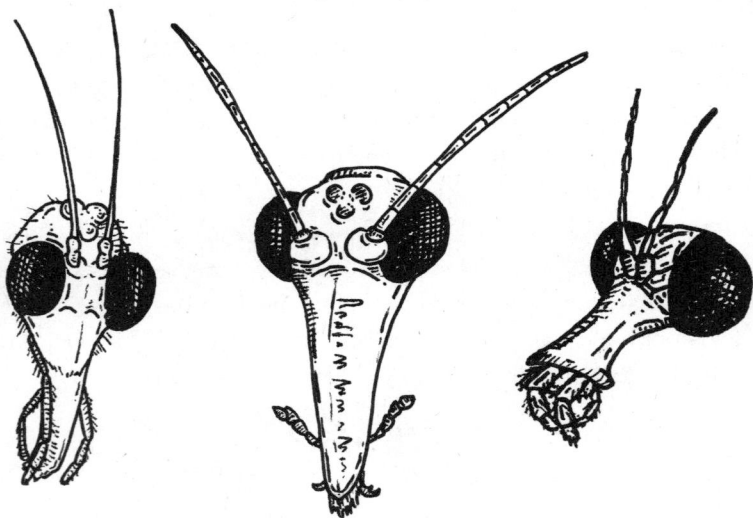

The largest differences between the faces can be noticed in the mouth parts. Crane fly (left), scorpion fly (middle), and hangerfly (right).

Hangerflies really look like crane flies. (Crane flies probably evolved from them). Hangerflies received that name because they hang on to the bottom of leaves with their little toes and don't fully chill on the leaf.

Unlike true flies (order Diptera), both scorpionflies and hangerflies have four wings instead of two. There are fewer species of hangerflies, around 150; sixty additional but now-extinct species appear in fossil records. These species are believed to have been important to pollination before bees evolved.

But hangerflies aren't pollinators these days. Similar to scorpionflies, males present their potential mates with a nuptial food gift. The larger the gift, the more likely the female will be to mate with them. Depending on when they lay eggs, the eggs may go into diapause and not hatch until the spring, meaning they may wait up to 290 days to hatch.

Adults are scavengers who will eat whatever is available to them. That is often plant matter, but hangerflies are not shy of eating animals. Sometimes males will pose as females to steal the nuptial gift of a dead bug from a male.

Hangerflies are also useful in forensic entomology. Unlike other bugs, they can be some of the first to show up after a death—only within the first two days. Entomologists use this behavior to determine how long it has been since an animal died.

(To this day, I've still never seen a hangerfly in real life, at least not that I've realized. I will be on the hunt for one this summer for sure. If anyone runs into one, send me a pic.)

How Human Culture Is Shaped by Scorpionflies and Hangerflies

In the video game *Fallout 4*, there are giant, irradiated stingwings who are really hard to beat and cause the player a lot of damage when they attack. They have the same physical features as a male scorpionfly (but significantly larger), but they also have a functioning stinger, which is definitely something that scorpionflies do not have in real life. (Who knows? Maybe that will become irradiated and evolve over hundreds of years to have stingers. Thankfully, for now, they may look intimidating but are harmless.)

The face of scorpionflies reminds me a lot of Garindan (also called "Long Snoot") in *Star Wars, Episode 4*. I'm unsure if these bugs were an inspiration for the character. But since antlions may have inspired the Sarlacc Pit Monster, who knows? Maybe bugs were a large influence on George Lucas!

Since there are so few species of hangerflies, I can find no reference to them in culture. They are often misidentified as craneflies so may appear in place of them in some media. The difference between them is hangerflies have four wings, and craneflies have two.

Quick Facts: Scorpionflies and Hangerflies

- While males do have something that looks like a stinger, it isn't one and is incapable of causing harm. Something inspired by these bugs appears in the video game *Fallout 4*, but that creature doesn't reflect real-life. Again, scorpionflies can't do that in real life.

- Hangerflies are called that because of how they delicately hang from vegetation instead of sitting atop a plant. It looks like they could fall off at any second.

- Male hangerflies bring a gift of food to the female before mating. The gift is often a meal, like a dead house fly. Scorpionflies offer a salivary gift (spit) after copulation.

- Forensic entomologists use scorpionflies to determine time of death since these bugs only show up on dead bodies within the first two days. If they are present, it is a sign that the body hasn't been dead for very long.

- Mecopterans are pretty far back on the evolutionary tree and entomologists believe that lepidopterans (moths and butterflies) and dipterans (flies) evolved from them.

CHAPTER 10: DOBSONFLIES, ALDERFLIES, AND FISHFLIES (ORDER MEGALOPTERA)

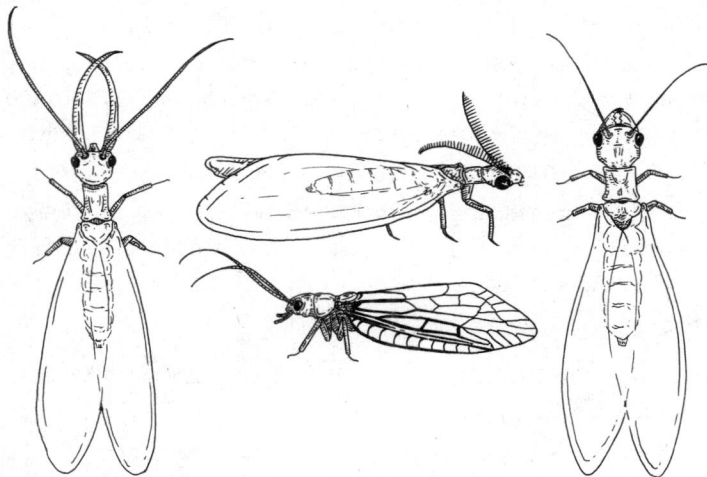

Dobsonfly male (left), dobsonfly female (right), alderfly (bottom middle), and fishfly (top middle)

*H*ere is yet another order with *fly* in the name, without being part of the order of true flies. Megalopterans have been around for a long time (at least 250 million years) but are still unknown or misunderstood by most people. These insects are aquatic as eggs and larvae, but terrestrial as adults. Similar to other insects, their life as adults can be very short. And all they do during that period is mate. Most male adults don't even eat.

Dobsonflies are more recognizable with their large jaws and huge size. Both male and female adults have large pincers, but the males' are extremely exaggerated. They can be about half the length of their three-inch body (four- to five-inch total in size). The females' mouthparts are shorter but still intimidating.

Fishfly males have interesting feathery antennae, looking similar to moths. The rest of the species in this order do not have feathery antennae. Fishflies and alderflies do not have huge mandibles and may only reach three inches in length.

Some of the largest aquatic larvae/terrestrial adult combinations in the world, dobsonflies mostly live in the Americas, with forty-three species living in North America. There are around three hundred identified species total. Other species live in Europe, the southern part of Africa, a few parts of Asia, and a few species in Australia.

They live statewide in Missouri, and I have never seen a dobsonfly, alderfly, or fishfly in real life. Not sure how that happened. The creek by my house was super polluted, so that may have something to do with it.

How Dobsonflies, Alderflies, and Fishflies Shape Our World

In their larval form, dobsonflies are called "hellgrammite." The name for them in the Ozarks of Missouri is "devil scratchers." The larvae have strong pincers that are used to catch small invertebrates and fish. While there is no clear history of how they got their name, my best guess would be because they are large with massive jaws. Encountering one in the water may seem hellish.

However, most of the time dobsonflies hang out under rocks, as they can do their best hunting from there while also avoiding predators. Fish love to eat them, so they stay hidden as much as possible. Once they are ready to molt into an adult (after about three years), they build a cocoon on land in which they overwinter to be born anew in the spring.

All the larvae for the different insects in this order look similar but have some differences. Hellgrammites are the largest and prefer fast-moving water. Fishfly larvae prefer slow-moving water and may be omnivorous. Alderfly larvae are very similar to fishfly larva: omnivorous or predatory, a preference for slow-moving water, and pupating on land (not in a cocoon, like dobsonfly larva).

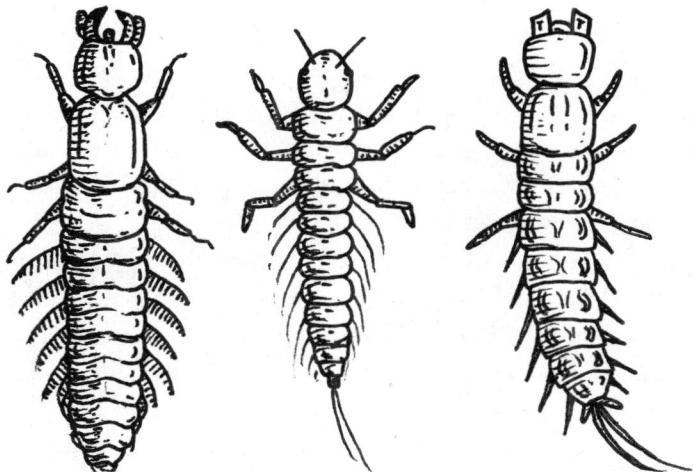

The larvae in this order are all much more predatory than the adults, who basically don't eat at all. Dobsonfly hellgramite (left), fishfly larva (middle), and alderfly larva (right).

As adults, male dobsonflies don't eat (they may drink water) and females may eat small amounts of fruit. Males bring females nuptial gifts and will fight other males for the chance to mate. Their mandibles are not used for biting but instead flipping their opponent over. Females will choose to mate with the champion of the wrestling match. Females can lay about 3,000 eggs at a time, and from birth to death, dobsonflies can live for around three to five years total. Of those five years, they only live three weeks in their adult form. The adults are very large and a good meal for many birds and mammals.

How Humans Can Interact with Dobsonflies, Alderflies, and Fishflies

The male dobsonfly's pincers are much more intimidating based on the sheer size; however, they can't really pinch you since they don't have enough leverage to create a serious bite.[29] The females, however, can pack a seriously painful pinch, so handle them with care (or at your own risk). Humans are unlikely to have interactions with either of them as dobsonflies are nocturnal and often hang out by water. They are attracted to light so, at night, they may fly around lamps. But, in general, they are clumsy fliers despite their large wings.

People who fish like using these insects as bait, although the practice isn't encouraged since faux versions exist. Many lures have been created to resemble them, and since most fish eyesight is fairly poor, they are easily fooled.

Megalopterans are not part of the EPT group (which includes caddisflies, mayflies, and stoneflies) used to gauge water pollution issues, but they are also known to be intolerant of pollutants. Lower populations in water where they were previously found is a signal for water issues in that area.

Some people in the US call mayflies "fishflies," but fishflies are a subfamily all their own. They look more like dobsonflies than mayflies. Regardless of which one you see, know that they may be a slight annoyance but are extremely important to water ecosystems. Seeing any and all of them is an indication that you live near healthy water. Fishflies also only live for three to fourteen days so their annoyance is short lived.

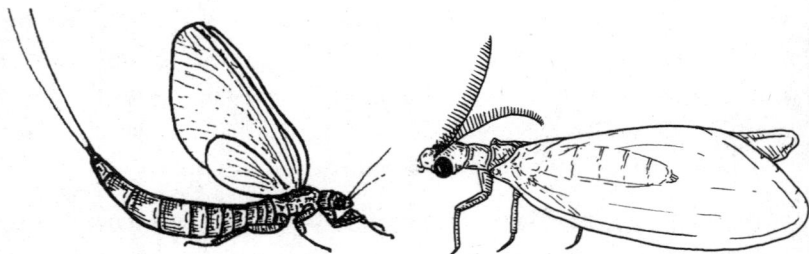

Fishflies and mayflies don't really look similar, but given the fleeting aspect of their life and their general location (water), mistaking one for the other makes sense.

29 Missouri Department of Conservation, "Hellgrammite," n.d., https://mdc.mo.gov/discover-nature/field-guide/hellgrammite.

How Human Culture Is Shaped by Dobsonflies, Alderflies, and Fishflies

Fishflies have inspired organizers and authors. In Baltimore, they have a fishfly festival that has, as of 2024, been going on for forty-six years.

Jeff Lemire, a famous comic book author and illustrator, created a comic called *Fishflies*, published by Image Comics. The story involves someone going through a strange metamorphosis, turning them into something they don't understand.

Despite their intense appearance, fishflies aren't found much in culture apart from how their larvae are described around the US. As mentioned above, in the Ozarks of Missouri, they call dobsonfly larvae "devil scratchers.' Other parts of the country call them "go-devils" or "grampus." *Grampus* is also a name that some people use to describe vinegaroons, a different bug in a different order. Their appearances are extremely different, so I'm unsure how the shared name came to be.

Quick Facts: Dobsonflies, Alderflies, and Fishflies

- They are not technically flies but can fly, though most often clumsily.
- As adults, they are terrestrial, but as larvae, they are aquatic. They all pupate on land.
- Along with caddisflies and mayflies, megalopterans are indicators of water health. A lack of their presence in waterways is alarming even though they are not part of the official EPT group. Some fishflies are a bit more tolerant of pollution.
- As adults, the males don't eat. Females may eat nectar or drink water but in extremely small amounts. Their MO is to mate and die (most often as food for other animals).
- There is a comic book, Jeff Lemire's *Fishflies*, reminiscent of Kafka's *Metamorphosis*, where a person is transforming into a fishfly.

CHAPTER 11: MAYFLIES (ORDER EPHEMEROPTERA)

S till another order that is called "fly" but ain't one. They are an order of insects whose name means "ephemeral" or "short-lived," which is an apt name for them. They live very fleeting lives and are often eaten before they expire.

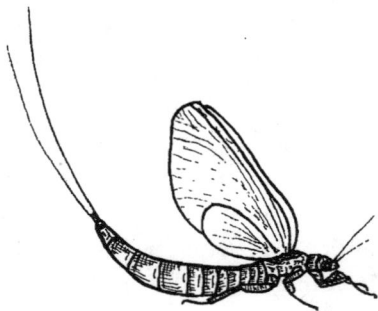

Mayfly adult

But even with this knowledge, as a bug-lover, I was horrified when I tried to get a few off my windshield once by using the wipers. It didn't move them; it just smashed them, and they were then much more difficult to get off of my window. That is a common interaction between people and mayflies. They are such delicate creatures with such interesting stances. Their alien-ness is highly apparent.

Nymphs are aquatic and adults are terrestrial, like dragonflies and caddisflies. The unique part of their lifecycle is when they first emerge from the water. They have a stage between emerging and full adulthood called their "subimago" stage, which is basically like them being a teenager.[30] They are not sexually mature in that stage but will molt once more to become an adult (the imago stage). They are the only insects who molt after developing functional wings.[31]

Unlike some of their aquatic/terrestrial cousins, most nymphs of mayflies species are vegetarians who eat algae and detritus. A few are omnivorous, but that is rare.

---

30 Shout out to Stefan for that comparison.
31 "Mayfly," Britannica, n.d., https://www.britannica.com/animal/mayfly#ref266454.

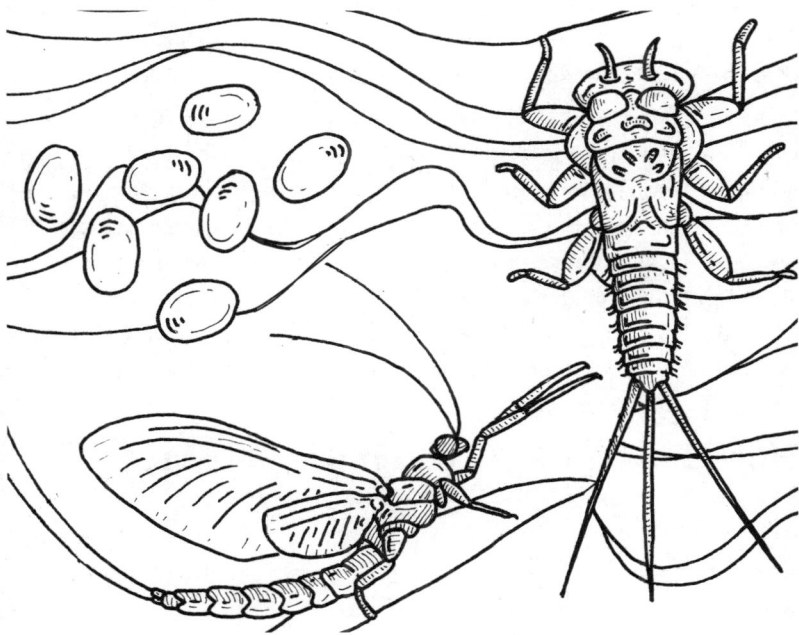

The lifecycle of a mayfly from egg, to larva, to adult. Notice the loss in one cerci between larvae and adults.

How Mayflies Shape Our World

Some adults only live one to two hours, and those that live the longest are only around for a few days. The whole purpose of their adult life is to mate. They don't eat as adults. Males have two penis-like appendages and the females have two sexual openings.

They are pretty recognizable with their wings resting up (like butterflies) and three long tail-like bristles coming out the back. Those are not stingers and mayflies are completely harmless. No venom, no stingers, no functioning mouth.

Mayflies do cause issues for humans sometimes, but not in ways one might expect. Often, of the 3,100 known species, large numbers emerge at once (kind of like cicadas), causing huge amounts to appear instantaneously. Sometimes the swarms are so large that meteorologists can see them on doppler radar.[32] Scientists believe these swarms are a defense strategy. Even if all the predators want to gorge on them, there will still be enough who successfully mate for next years' brood.

Some years, mayflies appear in such large numbers that they cover roads and windshields to the point of blocking visibility. They mate while they swarm, and their swarms can be overwhelming. News outlets in cities where their swarms are common will post information about how the city is trying to reduce swarms. Ohio reduced the number of streetlights on to deter swarming. These efforts don't magically make mayflies disappear, but they do reduce the chance of overwhelming numbers on car windshields. (And like I mentioned above, using your wipers on them is actually worse.)

How Humans Can Interact with Mayflies

Swarms are occurring less because of human activity. While that may seem nice for drivers, it isn't necessarily a good thing. Things like acidification of water, deforestation, general bad water quality, and pesticides all influence mayflies' numbers.

In 1953, Lake Erie, which has frequent mass emergences, previously had issues that caused a crash of almost extinction due to a severely unhealthy water. A decade later, water quality had been improved and since then, there has been a resurgence.[33] The diversity of mayflies was greatest during the Jurassic period, but many of those species are now extinct.

Mayflies are very sensitive to water cleanliness and are used as a macroinvertebrate to check water pollution. Because of this, they are part of EPT: they are the *E*, Ephemeroptera.

Larvae of the species are aquatic. They have gills and spend their time eating detritus while growing.

---

32 Missouri Department of Conservation, "Mayflies," n.d., https://mdc.mo.gov/discover-nature/field-guide/mayflies.

33 Resh and Cardé, eds., *Encyclopedia of Insects*.

Many species rely on mayfly larvae and adults for food: stoneflies, caddisflies, alderflies, dragonflies, water beetles, fish, birds, spiders, amphibians, marsupials, insectivorous mammals, you name it—if an animal eats bugs, there's a good chance they'll eventually eat mayflies in one form or another. Since mayflies are common in aquatic spaces, fly fishers use fishing lures that look like both the adults and larvae.

How Human Culture is Shaped by Mayflies

To humans, mayflies often symbolize the brevity of life. Mayflies have been featured in literature since *The Epic of Gilgamesh*, depending on the translation. Some works state that mayflies appear during a flood and are quickly gone.[34]

A TV show called *Mayflies* is based off of a book by Andrew O'Hagan and is about living with terminal cancer and realizing how quickly life can slip away from you.

Betty Buckley, famous Tony award winner, created a short animated film about a mayfly after seeing one fly overhead at an outdoor concert she attended. She liked the fact that the bug had no idea what else was happening; did the mayfly wonder if all the humans were clapping for it?

Finally, some cultures eat mayflies for food, like in Malawi where they use a paste made from mayflies to make cakes.[35]

34 Craig Macadam, "The Curious History of the Mayfly," The Freshwater Blog, May 16, 2011, https://freshwaterblog.net/2011/05/16/the-curious-history-of-the-mayfly/.

35 Peter M. Grant, "Mayflies as Food," Trends in Research in Ephemeroptera and Plecoptera (2001). https://doi:10.1007/978-1-4615-1257-8_14.

Quick Facts: Mayflies

- They live very fleeting lives (sometimes only one day) and in literature, often represent the fleeting nature of life. Perhaps unsurprisingly, they don't have mouths since eating isn't important to them in adulthood.
- Like cicadas, they can emerge in extremely large numbers, sometimes even causing visibility issues.
- When they are around, all the animals that eat them (of which there are many) are very happy and very full.
- Their larvae are aquatic and are another group that is important to determining water health. You want large numbers of their larvae to indicate clean water.
- The long wispy hairs out of their butts (cerci) are not stingers. They are used to detect vibrations and air currents. The cerci are harmless to humans.

CHAPTER 12: STONEFLIES (ORDER PLECOPTERA)

*A*s with many other bug orders, stoneflies are not true flies from the order Diptera. Their order, Plecoptera, means "folded wing" or "braided wing" in Greek since they fold two wings under the other when at rest. Even though they have wings, they aren't great fliers and prefer to walk from place to place. Because of this, certain species may inhabit a very small amount of space.

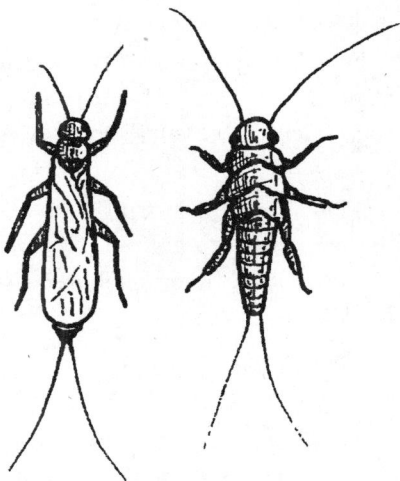

Stonefly adult (left) and nymph (right)

Adult stoneflies themselves can be fairly small (less than a half inch) to around two and a half inches.

So far, scientists have discovered around 3,500 plecopteran species worldwide—found everywhere but Antarctica. The earliest fossils date to 295 million years ago, but stoneflies looked much different then. Current species look more similar to fossils from at least 66 million years ago.

As adults, they have four wings and although they are terrestrial, they spend all their time close to water. By contrast, aquatic nymphs do not have wings and will spend their time in water with the other aquatic invertebrate nymphs like mayflies, dragonflies, and others.

How Stoneflies Shape Our World

Some species of stonefly have evolved to not have wings, but that is not a common feature. Similar to mayflies, they are mostly herbivorous/detritivorous and are important as a food source for animals. (There are always exceptions to the rule: some nymphs are predatory.)

Females lay upwards of 1,000 eggs in the water: some do it by dropping their eggs while flying, some set them in the water, some swim underwater completely to secure their eggs—it all depends on species.

They are the *P* in EPT to assess water quality, meaning they are very sensitive to pollution. Nymphs need highly oxygenated water, which they won't find in polluted rivers. Having them in your water is a good sign.

Their whole lifecycle takes anywhere from one to four years, spending most of those years being nymphs.[36]

As adults, they hide in rocks and crevices during the day. At night, they come out to mate during their week of life as fully grown stoneflies. Some eat, but it isn't common. More species only drink water.

In order to mate, males and females communicate by vibrating their abdomens on substrate to create calls. Females will sometimes refuse a male that approaches her by lifting her abdomen high into the air. Males mate with as many partners as they can and will sometimes even try to mate with other males, dead females, and nymphs. Some females in certain species may mate only once while other females may mate numerous times. Throughout the copulation, the males and females rub their antennae together as a form of communication.

How Humans Can Interact with Stoneflies

A specific wingless species of stonefly only lives in a small area in Victoria, Australia. Scientists tried to get it protected, but the

---

36 Just a reminder that nymph refers to insects that undergo incomplete metamorphosis, while larvae refers to insects that undergo complete metamorphosis.

government voted against it in 2024.[37] Entomologists are worried about the population since it has already declined eighty percent and is vulnerable to anthropogenic factors.

There are also two stoneflies in the US protected by the Endangered Species Act (ESA). The two species rely on cold water environments in glacier runoff. People worked tirelessly for ten years before they were officially protected under the act. It takes a serious amount of research and lobbying to get animals on the ESA.

Conservationists in Britain ran into similar issues with a native species called the Scarce Yellow Sally. They were able to create a successful captive breeding program in 2023 that greatly increased numbers. (But if the water quality or other anthropogenic factors get worse, they may see a decline in numbers again.)

The story in Britain is hopeful. So is the story of a group of fifth graders, who in 2024, made a case for stoneflies. A class of ten- and eleven-year-olds in Michigan tried to get the stonefly as the official state insect (it currently does not have one).[38] They chose the insect as their top contender because stoneflies are so integral to the health of Michigan waterways. Having younger folk advocate for insects is a good sign. It means that people are beginning to care for some previously unloved or overlooked creatures.

This order has been shaped by humans in a very interesting way. In 2024, a scientific paper was published that talked about how a species of stonefly had evolved a different color in response to human-driven factors. Factors like deforestation and fragmentation have caused non-toxic species of stonefly to change color to mimic toxic species of stonefly. Scientists believe this has happened as a result of being more vulnerable due to lack of coverage; their research is one of the first

37 Australian Associated Press, "Critically Endangered Stonefly with 'Remarkable' Lifespan Faces Extinction as Victorian Government Decision Condemned," *The Guardian*, March 12, 2024, https://www.theguardian.com/environment/2024/mar/12/critically-endangered-mount-donna-buang-wingless-stonefly-victoria-government-extinction.

38 Riley Kelley, "Fifth-Graders Say "Very Special' Stonefly Should Be State Insect," School News Network, https://www.schoolnewsnetwork.org/2024/03/11/fifth-graders-say-very-special-stonefly-should-be-state-insect/.

projects "that provides an absolute demonstration of human-induced evolution in the wild."[39]

How Culture Is Shaped by Stoneflies

Stoneflies are extremely important in fly fishing. There is a novel about fly fishing called *Yellow Stonefly* that is about the sport. Many fishing lures are made to look like nymphs and adults, as both are important food sources for fish. Because of stoneflies' vulnerability, it is not encouraged to use live specimens; instead, go for the reusable, look-alike lures.

Quick Facts: Stonefiles

- They are not true flies as they have four wings (instead of two) and are in their own order. Even though they are capable of flying, they are not very good at it.
- The adults are terrestrial or semiterrestrial, while the nymphs are exclusively aquatic.
- They are sensitive to anthropogenic factors like pollution and habitat loss. Some populations have declined by eighty percent over the last decade.
- They are an important indicator in water health and are used to determine how clean or dirty a body of freshwater is.
- They are popular in fly fishing, with many lures made to look like them.

NOTE: We have reached the end of the insect section and are now moving into the non-insect bug section. There are numerous insect orders that weren't covered in this book. Even those that were featured? Their chapters could contain significantly more information.

39 Steven Ni, Graham A. McCulloch, Gracie C. Kroos, Tania M. King, Ludovic Dutoit, Brodie J. Foster, Kahu Hema, Jennifer M. Jandt, Mei Peng, Peter K. Dearden, Jonathan M. Waters, "Human-Driven Evolution of Color in a Stonefly Mimic," Science 386, no. 6720 (2024), https://doi:10.1126/science.ado5331.

I hope what you've learned so far has inspired you to look into these orders more and to appreciate them, even if you often find them annoying. Now, we move on to non-insect bugs.

PART II: Non-Insect Bugs—Arachnids, Crustaceans, and Other Invertebrates

Many animals that people call bugs are not insects. None of the bugs in this section are part of the class Insecta. Most of the bugs in this list are in the phylum Arthropoda. That group includes insects, spiders, millipedes, aquatic species, and many more.

Some bugs do not belong to the phylum Arthropoda but are still invertebrates: slugs, snails, worms, etc.

In this section, all of the species have three things in common: their lack of a spine, their loose qualification of 'bug', and their lack of inclusion in the class Insecta.

Lacking a spine makes them sound unable to take on the challenges presented to them, but it is quite the opposite. Many arthropods existed before insects and most invertebrates have survived more than one mass extinction. The oldest fossil record of arthropods is from 450 million years ago (scorpion and millipede). The oldest invertebrate fossil record is from 3.4 BILLION years ago. Invertebrates have been the main species on earth since life existed.

And they still are—as 97 percent of all living animals. In this book, these bugs are all grouped together, but each one could easily have their own full books (and do). Not everything interesting or important about them can be covered here. Hopefully this section gives you a taste to do some more exploring.

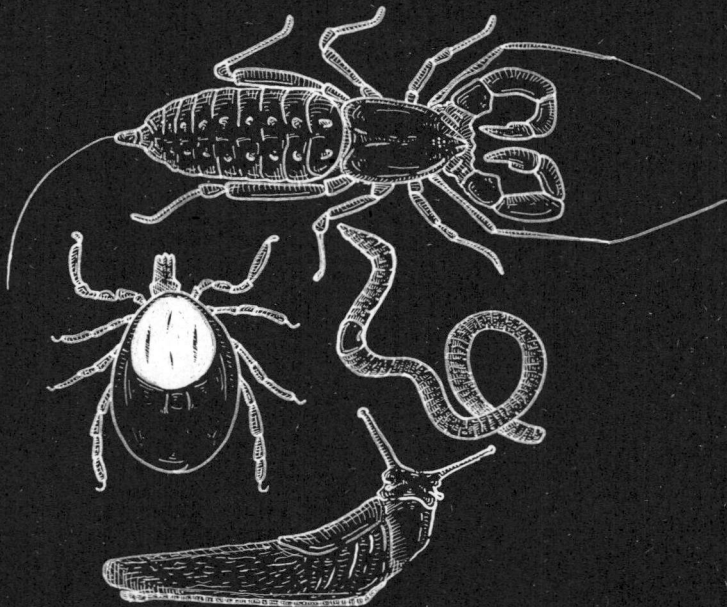

Vinegaroon (top), tick (middle left), earthworm (middle right), and slug (bottom)

- You may have heard that humans eat eight spiders a year: it's not true. We most likely eat zero spiders a year. We're too damn loud and big for spiders to be curious about our mouths. And we normally feel if something is crawling on our face.

- Contrary to popular belief, daddy long legs aren't hyper venomous. They do not have a venomous bite for humans— and their mouths are too small and their fangs too soft to pierce human skin.

- Putting nail polish on your chigger bites will not help force them out: By the time you start itching, the bugs are already gone. They do not burrow into or live in your skin but feel at the surface.

- Spider females don't always eat spider males after mating: Most of that happens in laboratories. In the wild, it happens infrequently.
- Not all ticks carry Lyme disease. Only certain species carry Lyme disease, though ticks are known to carry many other ailments.

Class Arachnida is well represented in this section: scorpions, spiders, ticks, mites, pseudoscorpions, whip scorpions, vinegaroons, and camel spiders. They are related to insects but are in the class Arachnida (all of which go by the common name Arachnids).

But arachnids' bodies are slightly different from insects. They have two body parts: the cephalothorax (kind of a head/thorax combo) and the abdomen. They also have eight legs and many have what are called pedipalps, which can sometimes be long enough to look like a fifth pair of legs. They may also have pronounced chelicerae, which are the appendages they use to help them eat (or where some have their venom).

Other bugs can be found in the phylum Gastropoda (snails/slugs), class Malacostraca (crawfish/pillbugs), and phylum Annelida (worms), to name a few. The body parts in these bugs vary greatly, yet, yes, are all non-spined animals.

CHAPTER 13: SCORPIONS (ORDER SCORPIONES)

*T*rue scorpions are an ancient group of creatures that inspire horror and fear in most people. They are the oldest arthropods, dating back 450 million years. Their body structure is still fairly similar now to how it was then, to the surprise of scholars. They assumed things would have changed a bit more over hundreds of millions of years.

But if it ain't broke, don't fix it. Scorpions have been around the entire time humans have existed, which explains why they are discussed in *The Mayan Book of the Dawn of Life* or why there is a yoga pose named after the scorpion.

Top down view of a scorpion. The general rule is the larger the claws, the less venomous the scorpion. Probably not worth testing if you can avoid it.

How Scorpions Shape Our World

Currently, there are about 2,000 known species of scorpions. Some are tiny (approximately 1/4 of an inch), with the largest being around nine inches long. In scorpions, the pedipalps are actually where their claws are. Unlike other arthropods, they don't seem to have issues with wasp or fly parasitoids.

Their stinger is unique to true scorpions. That is what they use to subdue their prey. Since they are incapable of eating solid food, they liquify the inside of their victims with an enzyme that breaks down the body before they suck it up. Most species only need to eat every two to six weeks, depending on food availability. So they can eat one cricket and be good for weeks. Not all go for crickets however; some have been recorded eating centipedes, other scorpions, lizards, and small snakes.[40] They are able to eat so infrequently because of how little they move. Their hunting style is "sit and wait."

Apart from what makes them scary, scorpions actually have surprisingly sweet sex lives and parental care. Many species have a sophisticated sex dance that happens between the male and female. First, they grab each other below the claws. Then, the male lays a spermatophore and dances the female over it. This dance is officially called the *promenade a deux*, French for "dance of two." The dance can last a few minutes or a few hours, depending on species. Some types

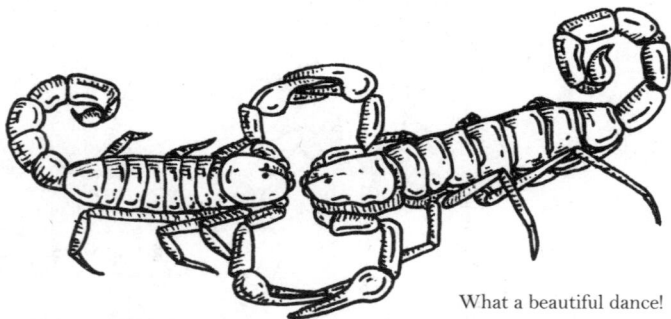

What a beautiful dance!

---

40 NOTE: Apparently, they can be pretty picky and thus avoid animals like stink bugs because of their unpleasant odor.

add in other parts of the dance, like rubbing their mouthparts together (called a kiss) or putting their tails over their bodies to touch overhead.

After the sex dance is complete, gestation takes anywhere from two months to almost a year. Unlike many arthropods, scorpions have a pretty well-documented maternal care structure. Females carry their babies as embryos and birth a dozen to over two hundred live young (depending on the species).

After that, the little ones climb onto her back—in a special place made for them there, where the mother helps feed them until they can feed themselves. They will hang out with her for quite some time afterwards, between a week to a month depending on the species.

How Humans Can Interact with Scorpions

People rarely see scorpions because most prefer to hide under rocks or in their burrows in natural areas. The majority of species can be found in arid or temperate areas, with some on the fringes who live in intertidal/beach habitats or the one species in California that can be found at 10,000 feet.

Most species can be folded into one of two categories: bark and ground. Bark scorpions like to hang out on trees, so should they get into your house, they are more likely to be found on the walls or the ceiling.[41] Ground species create burrows where they spend most of their time. They like their burrows to be tight and just big enough to squeeze themselves in.

Many species, both bark and ground, glow in UV light. In Arizona, you can go out into the desert for a night hike, bring a black light, and see scorpions from fifteen feet away (which may be the preferable distance for some people). Some biologists believe that scorpions may

---

41 NOTE: They are unable to climb completely flat surfaces, like non-matte plastic, so keep that in mind when buying garden furniture.

be fluorescent to sense the phases of the moon, while others believe it may be to detect light.[42, 43]

Most fears associated with scorpions are tied to their stingers and secretive nature. Only about twenty species worldwide (out of 2,000) have the possibility of being deadly to humans, and only one of those species lives in the US: the Arizona bark scorpion. Their venom is really only meant to defend themselves or subdue their meal. Overall, stings are down around the world because of antivenins, public knowledge, and less sightings in human dwellings.

Antivenins are available, especially in places where stings are most common (like Brazil and Mexico). To make the necessary meds, scorpions are milked by using electrical stimulation, stunning them for a bit. This process can be dangerous for humans as the scorpion is alive and can strike. Some Moroccan students made a robot to do the job, but it has yet to become a widely used tool.[44]

Some cities are experiencing increased numbers of bites because of encroachment on scorpions habitats or lack of natural predators. This is the case in southeastern Brazil with the yellow scorpion. Researchers in 2020 found that a toad is a "voracious predator" of the yellow scorpion and has very little reaction to its venom if stung in the process.[45] They think that these toads are the future of keeping yellow scorpion numbers down and reducing human interaction, since the sting can be fatal to humans.

42 "Zoology: Scorpions Glow to Sense," *Nature* 468, no. 734 (2010), https://doi.org/10.1038/468734c

43 Carl T. Kloock, Abraham Kubli, Ricco Reynolds. "Ultraviolet Light Detection: A function of Scorpion Fluorescence," *The Journal of Arachnology* 38, no. 3 (2010), https://doi:10.2307/20798577.

44 Eduardo Gonzalez-Ponce 1, Sofia Rodríguez-Rangel, Sofia Rodríguez-Rangel,Raymundo Martinez, Adrian Alvarado, Estela Ruiz-Baca,Pablo Miranda, Jorge E. Sánchez-Rodríguez, and Angelica Lopez-Rodriguez, "Scorpions, Science and Folklore in Durango City," *Diversity* 15, no. 6 (2023), https://doi.org/10.3390/d15060743.

45 Carlos Jared, César Alexandre, Pedro Luiz Mailho-Fontana, Daniel Carvalho Pimenta, Edmund D Brodie Jr., Marta Maria Antoniazzi "Toads Prey upon Scorpions and Are Resistant to Their Venom: A Biological and Ecological Approach to Scorpionism," *Toxicon* 178 (2020), https://doi:10.1016/j.toxicon.2020.02.013

Ultimately, scorpions are misunderstood. Keeping your distance is always encouraged, which goes for any creature you don't know much about.

Settling around cautious appreciation doesn't sound like a bad place to be. Scorpions are walking fossils and have much more experience on the earth than we do. They are important to the ecosystem, so learning how to best live with them is necessary.

How Human Culture Is Shaped by Scorpions

There are few animals, including other scorpions, that may be resistant to scorpion stings: mongoose, grasshopper mouse, long-eared desert bat, and some toad species. Other species still eat them, even if they are at risk of being stung, including owls, various bat species, coyotes, and humans.

Scorpions do not want to strike, so they warn you before they resort to that: they use stridulation as a warning noise, attack with claws first, squirt a venom upwards of three feet, and lose their tail. If they lose their tail, it does not grow back, meaning they can't sting or poop but can still live for about eight months after.[46] They have a 360-degree view like flies, so they know when you're coming.

Humans in Africa, Mexico, and Asia eat scorpions in various ways. In Vietnam, they are an integral ingredient in snake wine (some researchers call it scorpion wine). It is exactly what you think it is: venomous snakes and scorpions soaked in grain alcohol. It is apparently good for arthritis.[47] In Mexico, there are some distillers that use scorpions in making Mescal. They are also included in novelty

46 Camilo I. Mattoni, Solimary García-Hernández, Ricardo Botero-Trujillo, José A. Ochoa, Andrés A. Ojanguren-Affilastro, Ricardo Pinto-da-Rocha, Lorenzo Prendini, "Scorpion Sheds 'Tail' to Escape: Consequences and Implications of Autotomy in Scorpions (Buthidae: Ananteris)," PloS One 10, no. 1, https://doi.org/10.1371/journal.pone.0116639.

47 Dirk W. Lachenmeier, Pham Thi Hoang Anh, Svetlana Popova, Jürgen Rehm, "The Quality of Alcohol Products in Vietnam and Its Implications for Public Health," *International Journal of Environmental Research and Public Health* 6, no. 8 (2009), https://doi.org/10.3390/ijerph6082090.

suckers you buy at gas stations. Heating their tails negates the toxins' effects, so no worries there.

In addition to being food and creating antivenin, scorpions are important to humans in other ways. Their venom is used in both traditional and modern medicine. Researchers are trying to see if venom could be used for anticancer drugs.[48] In traditional medicine, people will crush a whole scorpion into powder and use it as an anti-epilepsy drug. In Mexico, some people will use crushed scorpions to treat scorpion stings.

Scorpions are also commonly kept as pets, as they are interesting and relatively easy to care for since they need minimal food and water. (Excessive activity in a captive scorpion is concerning, since they mostly do nothing.) They live one to five years in captivity. Unfortunately, many threatened species are taken for the pet trade, so if you want to keep one as a pet, be sure to do your research.

As part of the Big Bug trend of the 50s and 60s, *The Black Scorpion* was a 1957 Mexican-American movie about giant prehistoric scorpions that were released due to volcanic activity.

In *Fallout 4*, players encounter Radscorpions. They burrow underground and spring out right next to you. It exaggerates their size, aggression, and burrowing abilities. They are very hard to beat, and you have to go for that stinger first or they will take you out.

Scorpions were common in ancient Islamic art. In astrology, there is the Scorpio sign, and the Egyptian Goddess Serket was often portrayed as a scorpion.

More recently, scorpions have been used to name many deadly weapons, as well as give the name to a string of movies about the Scorpion King (with some really terrible CGI).

---

48 Shirin Ahmadi, Julius M. Knerr, Lídia Argemi, Karla C. F. Bordon, Manuela B. Pucca, Felipe A. Cerni, Eliane C. Arantes, Figen Çalı□kan, Andreas H. Laustsen, "Scorpion Venom: Detriments and Benefits," *Biomedicines* 8, no. 5 (2020), https://doi.org/10.3390/biomedicines8050118.

The band Scorpion was a hair-metal band from the 80s who definitely thought they were badass. Actual badass scorpion characters appear in *Mortal Combat* and the Marvel comics.

QUICK FACTS: SCORPIONS

- Scorpions are one of the oldest arthropods. They first showed up in fossil records from 450 million years ago.
- They don't need much food to survive and can go six weeks between meals. Humans are never on their menu, but in the US, if you get stung by a venomous one, seek medical care.
- Their mating rituals are elaborate and dubbed by scientists as *promenade a deux* since they basically dance with each other. Some species will touch their tails together over their head.
- Female scorpions are caring parents who will let their children hang out on their back until they are ready to go off on their own.
- In culture, scorpions often represent some sort of badass, like a villain in comic books or something similar. The scorpions in *Fallout 4* are irradiated versions of the real-life ones. They are much different from the current scorpions and do not have any actual behaviors of real ones. Don't let those ones scare you away from appreciating scorpions!

Chapter 14: Spiders (Order Araneae)

Compared to scorpions, spiders have a huge number of known species: over 51,000, with tens of thousands of yet-to-be-identified species. Spiders are often more conspicuous than scorpions (though there are always exceptions to the rule).

Quite a few spider species have adapted well to human dwellings. As a member of the Arachnid family, they also have a cephalothorax and abdomen, as well as four pairs of legs with pedipalps (that can be long, skinny, lobed—all different kinds).

Most spiders have eight eyes, though some can have fewer, and a few cave-dwelling species have none at all. Many spiders have poor eyesight and rely heavily on their palps and front legs to figure out what is around them.

Jumping spiders are a different story. They can recognize themselves in a mirror and have much better eyesight than most spiders. It's part of what makes them such successful hunters.

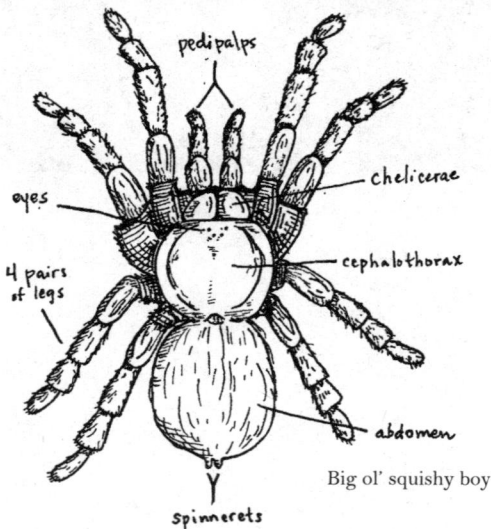

pedipalps

chelicerae

eyes

cephalothorax

4 pairs of legs

abdomen

spinnerets

Big ol' squishy boy

How Spiders Shape Our World

Silk is intimately tied to every stage of life for spiders. Some arthropods use silk at different times in their developmental stages (i.e. caterpillars, adult ants, caddisfly larvae), but spiders are the only arthropods to use it throughout their entire development.

The silk travels through upward of six hundred ducts on the inside of a spider's body before exiting the spinnerets. Spiders then use silk for quite a few things:

- The most iconic spider-related thing made with silk, webs come in varying shapes and sizes. Orb weavers create the classic spider web, but many webs don't look like that. Some are more like webby mats. In laboratory studies, spiders have been given drugs and caffeine to see how it affects them. The experiments resulted in irregular webs.

- Sometimes, spiders use smaller amounts of silk to help keep prey from breaking free from the web. Some specialized predators dangle a single line of silk to catch a meal.

- Spiders are agile, but they always have a helpline should they need one: They create draglines to stop them from falling. Draglines can stretch up to thirty percent beyond their original length. It's like having someone belay for you all the time.

- Shortly after the final molt to maturity, a male spider makes a small web (sometimes just one or a few strands of silk). In that silk, he wraps up a little sperm to carry around with him to give away. He then places the tip of his palp into the sperm, transferring the sperm into his palps on his face. He stores the sperm there while waiting for a mate.

- Females use silk to make a place for their eggs or young. Sometimes, this looks like a small ball that she attaches to her abdomen or spinnerets. Sometimes, it is more like a silk hammock for her eggs to hang out in.

- Some species of newly hatched spiders (and sometimes adults) will climb to the highest possible point (even if that is just a blade of grass) and create a "balloon" of sorts, allowing the wind to carry them away to other locations. This is how some spiders show up in previously unoccupied spaces.

In their web, a spider will generally respond to the vibration of an insect caught in the threads by shaking the web, causing the insect to become more firmly stuck and allowing the spider to locate the position of the insect. The spider may then run to the caught animal (often an insect) and wrap it up, much more so if the prey is large and/or strong.

Not all spiders have webs. Some like to wander, like the wolf spider. They got their name from their wandering nature and because of their expert hunting skills. They use their silk for other things (like egg sacs and draglines).

Hunting can look different for different species. The genus *Portia* is a group of fairly intelligent spiders who only eat other spiders. They have been observed outflanking and luring their prey.

Some have both a distinct nest and feeding style. A species only found in New Zealand lives in rocks on the beach between high and low tide areas. When the tide comes in, it retreats to a little silken tube and blocks the entrance with silk. The tube is completely submerged and the spider won't come out until the tide retreats to find food.

Mating is also tied to their silk in many ways. Courtship looks different in different species but some males can pluck the web of a female to get her attention. They'll carefully strum, sometimes for hours. If they don't use webs, some vibrate in litter to create noise.

Some jumping spiders, like the peacock jumping spider, even do big showy courtship dances by waving legs and abdomens with bright colors and patterns. (If you are looking for a cute video to watch, this is a great place to start.[49])

49 NOTE: If you don't want to watch the peacock jumping spider mating dance, I highly recommend you look into stories and videos about how some tarantulas have frog pets. It will

Sometimes the females go for it, sometimes they don't. After they're done mating, sometimes the female eats the male. This is true of all spider species but rarely happens. Even species like the black widow, which is notorious for doing so, rarely eat their partner. Most of the examples of females eating males are in laboratory settings. Laboratory conditions seem to do interesting things to bugs.

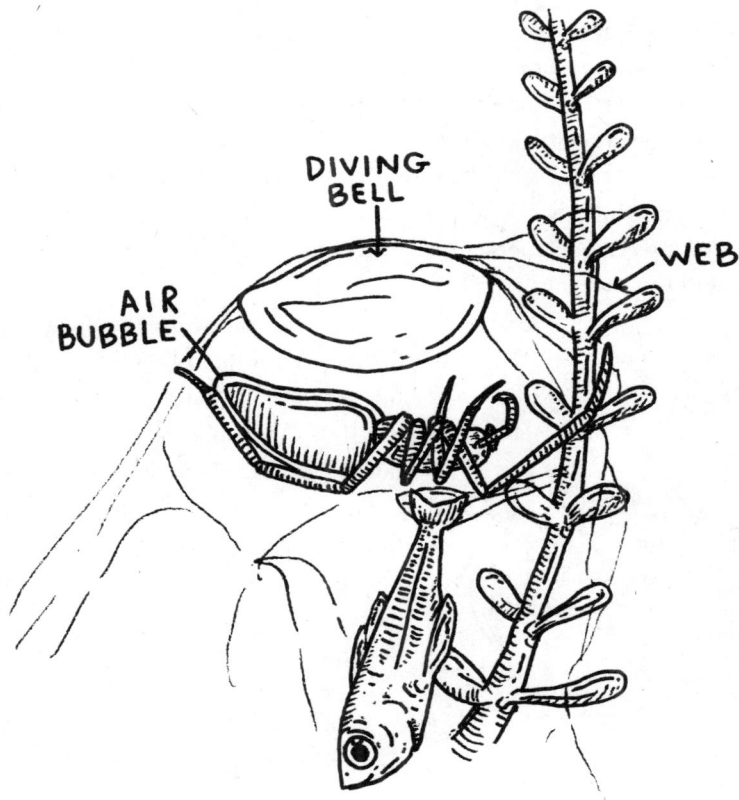

What a fascinating evolutionary strategy!

brighten your day or at least give you something fun to tell your friends.

Many females carry their egg sacs around with them (like wolf spiders, since they are nomadic) or by watching them in their nest (like black widows). Once the young have hatched, a few species carry them around on their abdomen until they are developed enough to go out on their own. Some even feed their young. In the biggest sacrifice of all, two species are known to give their own body to their young to eat, called matriphagy.

If they don't give themselves up for matriphagy, some spiders live for anywhere from one to thirty years (particularly tarantulas in captivity). They continue to molt throughout all of their life stages. Sometimes what ends up killing them is that they get stuck in a molt at an advanced age. Species that live for years like the black widow will overwinter, which is similar to hibernating. They find a warm, safe space to lower their metabolism and remain relatively still until the weather gets warmer.

Look up these guys in color. The camouflage is unbelievable.

One of the ways they are able to live so long and be such successful hunters is their mimicry and defense. Crab spiders blend in with flowers by being the same color (like the orchid mantis). Some purposefully look like poop because it is attractive to certain insects. Some spiders have the ability to change color depending on the tree they are in. A species of jumping spider looks like an ant, with its pinched abdomen. This allows it to move around closer than usual to prey.

Apart from venom and mimicry, some spiders like tarantulas have small hairs on their abdomen that they can shoot as a warning. If they do it often enough, it looks like they have a bald butt.

How Humans Can Interact with Spiders

Of all the topics that surround spiders, two things are talked about most often: venom and silk. Venomousness varies drastically depending on the species. Many species do not have venom that is harmful to humans and some don't even have fangs big enough to break human skin. Common spiders, like wolf spiders or jumping spiders, are not known to bite humans. Even if they did, their bite isn't problematic. Chances are they will bite you only if you are actively smashing them and they can angle themselves to bite you, a situation that will inevitably kill them, so they're trying to get their last word in. Most spiders have venom that is specifically designed for use against invertebrates. Only a very few have been found to cause potentially serious issues for vertebrates including humans. In general, their venom is used to take down prey, not humans. Two species in the US have been known to have medically significant bites, which are the black widow and brown recluse.

Venom affects one of two things: the nervous system or muscles. Full envenomation of a human is rare due to many factors like how recently the spider had a meal (meaning they recently used their venom for that), the age of spider (the older, the less potent the bite and the less likely to be able to full bite) and the angle (with tiny mouths, they need to be perfectly lined up). Most people, even if bitten by the most venomous spider, will recover in twenty-four to forty-eight hours.

It is important to mention that necrotic tissue damage is not guaranteed in brown recluse bites. Most healthy adults recover without medical intervention. You do not want to get bit, but you will probably be okay if you do.

My mom had a brown recluse that fell into her boot at work. She didn't feel it bite her initially (which is common) but then had a large, purple bruise-looking circle on her foot for weeks. Eventually, there was some shooting pain and discomfort, but she didn't need anything from the doctor. Again, unpleasant but no lasting issues.

Another family member who had serious health issues was bitten by a brown recluse and had a necrotic wound for months. These types of responses are more likely to happen when a person is unable to fully take care of themselves. So far, brown recluses do not live farther west than Kansas.

Black widows are widespread but prefer to be outside (compared to brown recluses, who are often found indoors in rarely-used blankets, shoes, coats, etc.). For being the most poisonous spider in the US, black widows' bites also have a very low fatality rate. Going to the doctor is recommended to be sure.

Just mentioning spiders can create visceral responses from people. Arachnophobia is very real but some researchers believe it may be increased by misinformation. A study in 2022 found that forty-seven percent of articles about spiders contained errors and forty-three percent were sensationalist.[50]

Arachnophobia technically includes scorpions and ticks but is most often associated with spiders. The feelings or phobia may be a mixture of evolutionary response and cultural factors. It is most common in European countries, and much lower in countries where people eat spiders and have positive representations of them in their culture.

In fifteenth century Italy, a wolf spider[51] was reportedly responsible for a dancing mania that overtook the town. The hysteria caused by the bite was only purged by giving in to dancing. The dance "tarantella" is believed to have come from this experience. Historians don't think the wolf spider was to blame since their bite doesn't cause

50 Stefano Mammola, Jagoba Malumbres-Olarte, Valeria Arabesky, Diego Alejandro Barrales-Alcalá, Aimee Lynn Barrion-Dupo, Marco Antonio Benamú, Tharina L Bird, Maria Bogomolova, Pedro Cardoso, Maria Chatzaki, Ren-Chung Cheng, Tien-Ai Chu, Leticia M. Classen-Rodríguez, Iva Cupic, Naufal Urfi Dhiya'ulhaq, André-Philippe Drapeau Picard, Hisham K El-Hennawy, Mert Elverici, Caroline S. Fukushima, Zeana Ganem . . . Catherine Scott, "The Global Spread of Misinformation on Spiders," Current Biology 32, no. 16 (2022), https://doi: 10.1016/j.cub.2022.07.026.

51 Their scientific name is Lycosa tarantula, but they are not tarantulas. Confusing.

issues (and they rarely bite). Instead, they think that it was actually a push against more restrictive social rules. Still doesn't put spiders in a great light.

Even for those with arachnophobia, spider silk still garners interest as a material that could be useful for humans. Both males and females can produce silk. It's incredibly strong for how fine it is and is resistant to decay; it has so many properties that could be useful to us. Scientists have been trying to produce it in a lab since the 1960s. People have been able to create spider silk in the lab in many ways, but none of them are yet plausible for mass production. Recently, some were able to create it using E. coli and certain genes from a spider. The most straightforward way to get spider silk in the lab is by "milking" spiders. To milk a spider, you have to knock them out, secure them down, and basically pull the silk out of their spinnerets.[52]

A large piece of clothing was made using silk from the golden orb weaver. The problem was that it took millions of spiders' silk and numerous craft folk an extremely long time to make it. Not great for Target t-shirts. In 2023, scientists at the University of Wyoming "successfully modified the genetic makeup of a new breed of goats that contain golden orb spider genes, allowing them to produce milk laden with spider silk proteins,"[53] meaning that they can extract spider silk proteins but I'm unsure if that is the same as the silk itself. A Japanese company, Spiber, has had some success mass producing silk without needing to milk spiders. They created clothing and jackets with their fabricated spider silk, so there may be more spider silk clothing, as well as Kevlar, ropes, and bandages, in our future.

How Human Culture Is Shaped by Spiders

Instead of having claws on their pedipalps, many spiders use them and their first pair of legs to "taste" what's around them: they have

52 David Willard, " 'Milking' Brown Recluse Spiders for Silk," Phys.org, March 10, 2014, https://phys.org/news/2014-03-brown-recluse-spiders-silk.html.
53 Dave Dempsey, "University of Wyoming Molecular Biologists Breed Spider Silk—Producing Goats," *Discovery Scientific Solutions*, October 2, 2010, https://www. discoveryscientificsolutions.com/item/31.

hairs that allow them to feel, smell, and determine. Their legs are basically moved with hydraulic pressure; it raises their heart rate a lot to use that pressure so they often travel in short bursts, which can be unsettling to some people. They only lift one or two legs at a time, never more than that. Because of how their legs work, researchers have tried to use dead spiders to grab small objects in the lab, in what has been dubbed "necrobiotics."

In 2022, a research team of biologists found that the dead spiders could pick up more than 130 percent of their own body weight and last through 1,000 open-close cycles. They believe that this tech could be used in the field to collect specimens like tiny insects since the spiders are naturally camouflaged.[54]

One unique species, the freshwater spider, is the only completely aquatic spider, even though it can't technically breathe underwater. It forms an air bubble on its abdomen to allow for breathing. It then leaves a pheromone-soaked silk dragline for males to follow. If she mates, the female lays her eggs and raises her young in a small diving bell underwater, held in place by silk.

In Cambodia, tarantulas are considered a delicacy, but those irritating hairs on their abdomen should not be consumed so they are cleared off before being fried and eaten.

Not many places eat spiders, but they are found in culture in numerous other ways. Arachnology, the study of spiders, was named after a Greek goddess. A young woman named Arachne challenged the goddess Athena to a spinning contest. Athena created a wonderful tapestry, but Arachne's was better. Athena was pissed, so she turned Arachne into a spider, condemning her to a life of web spinning.

In Cherokee folklore, it was a spider that brought fire to the people. Diné folklore has the Spider Woman (Na ashje'ii 'Asdzáá), who came from the "first world" and taught the women how to weave. According

54 Te Faye Yap, Zhen Liu, Anoop Rajappan, Trevor J. Shimokusu, and Daniel J. Preston, "Necrobotics: Biotic Materials as Ready☐to☐Use Actuators," *Advanced Science* 9, no. 29 (2022), https://doi: 10.1002/advs.202201174.

to Pueblo legend, Spider Woman was at the core of creation. Sioux peoples use the "dream catcher," spun by a spider, to capture the good dreams of life. Spiders are also prominent in African culture, as illustrated in the well-known stories of Anansi, the Trickster. Scottish legend tells of King Robert the Bruce, whose observations of a spider inspired him to persevere, going on to conquer the English. (If at first you don't succeed, try, try again.)

In the present day, spiders show up in our movies, music, and comics.

A spider-like kaiju appears in a Godzilla film (*Son of Godzilla*, 1967), as well as modern "big bug" films like *Eight Legged Freaks* (2002), who become large due to nuclear waste.

We all surf the World Wide Web, a nod to the spider's iconic silk structure.

You can see two very different representations of spiders in Roald Dahl's *James and the Giant Peach* (1996) and Pixar's *A Bug's Life* (1998).

Rick and Morty, the TV show, has a spider subplot in season 2, episode 6, in which humans and giant telepathic spiders come to a truce to live together in peace.

Spiders also show up in well-known kids' songs and nursery rhymes like "Itsy Bitsy Spider" and "Little Miss Muffet," as well as adult songs, like "Boris the Spider" (by the Who) and "Spiderwebs" (by No Doubt).

In *Zelda* games, you have to kill Skulltulas (a type of spider with a skull-shaped body)

These creatures appear in art, both classic and modern, like in the spider sculptures of artist Louise Bourgeois (1994 series).

Two of the most famous spiders of the last twenty years are Shelob and Aragog, giant spiders that appear in *Lord of the Rings* and *Harry Potter*. Neither are presented very positively. Shelob even has a large stinger (which does not exist in real-world spiders).

Spider-Man is probably the most famous spider-themed character in recent history. He is a beloved comic book hero who bears the name *spider* and shares some similarities. He even shoots silk from his wrists.

In some versions, he has to manufacture the webs and web shooters, and in others, he naturally produces it. He has "spider sense," which doesn't exist in real-world spiders, especially since some of them have very poor eyesight. He also doesn't have eight legs. Though, sometimes, he has them as part of a suit, or in the arcs he turns into Man-Spider. When he is Man-Spider, he has many more spider features.

Even with so little in common with an actual spider, having Spider-Man bear that title is good for generating positive feelings towards spiders. He, along with his other invertebrate-named pals, can help create more optimistic views of these often negatively-portrayed arthropods.

In the *Spider-Verse* animated film franchise, there are thousands upon thousands of different spider-people that fans love. Let's keep it going and growing.

Also, there is a very cute comic created by Donny Cates and Chip Zdarsky titled Peter *Talks to a Spider*. It is worth a read and addresses how little his character shares with real spiders.

Quick Facts: Spiders

- Spider-Man may be the most famous spider character but doesn't share a lot with the real-world counterparts. Still, having a beloved spider can do great things for human-spider relations.

- Along with often having eight eyes, spiders have lots of cool features: eight legs, pedipalps that almost look like a fifth pair of legs (for feeling their way around), chelicera (where fangs are, if they have them), an abdomen, and a cephalothorax (head and thorax in one basically).

- Silk is extremely important to spiders and humans. Spiders use it for a myriad of things: webs, eating, moving around, etc. Humans hope to harness it to create strong fabrics. Strides are being made in that field, but there is still a lot of work to do.

- Spiders are important in folklore from peoples across the world, from Anansi in Africa to the Spider Woman among Diné (Navajo peoples).

- Some spiders, like the wolf spider, are known for carrying their egg sacs with them and then carrying around their babies once they hatch. Some species perform matricide, giving themselves as a meal to their children in their last act as a mom.

CHAPTER 15: TICKS AND MITES (ORDER ACARI)

*T*hese two types of arachnids may be grouped together in an order, but they lead very different lives. One of them (ticks) is almost universally hated. The other (mites) are rarely thought about. Both however have extremely diverse habitats and can be found on every continent, including Antarctica. Their lifecycles and feeding habits are drastically different (in most species).

So how are they related? They both have similar bodies which include what looks like a single body part (no separated or noticeable abdomen). They also have no head, just sucking mouthparts. All ticks use those mouthparts for blood, whereas mites have more varied diets.

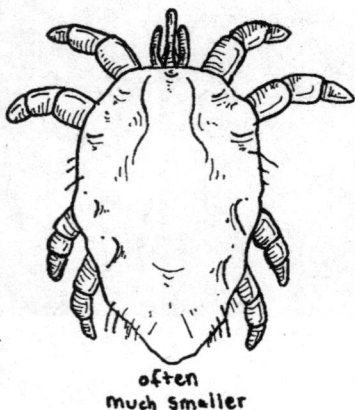

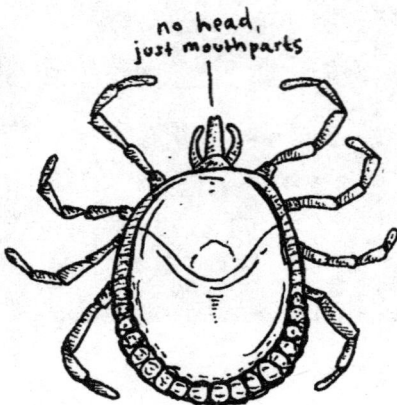

often much smaller

no head, just mouthparts

In this image, ticks and mites are similar in size. In reality, ticks are often larger than mites.

TICKS

Anyone who goes hiking in the Midwest or fully-foliaged forest during summer has had experiences with ticks. My mom once sat on a log to rest, and when she stood up, her back was covered in hundreds of small ticks. In Girl Scouts, we would have contests to see who had the most ticks after a day in the woods, which looking back on it, was a bad idea. Thankfully, there are bug sprays and clothing choices that greatly reduce, maybe to zero, the number of ticks you get.

Ticks' habitats vary greatly. They can be found at high elevations in dry climates or found on penguins in Antarctica.

Ticks are an arthropod that is difficult to talk up since most of what they do is not great. After reading this chapter, you may still not like them, but you may be able to appreciate their adaptations.

How Ticks Shape Our World

Around 650 species of tick have been identified, and fossilized amber tells us that they have been around for close to 100 million years. Many scientists separate them into two groups: soft-bodied and hard-bodied.

These little weirdos don't have a head, which is almost unbelievable, and instead have large mouthparts at the front of their body where the head normally is (their brain is located further back in the cephalothorax). They need to feed on blood to grow and many aren't picky about their host. Mammals, reptiles, birds, anything with blood—all of these are on atick's menu.

Some species are host-specific, like some bird ticks, while others are generalist feeders. If blood isn't available, they can survive for long periods of time waiting for another host to come by. Some have been observed living for upwards of a decade waiting for their next meal. They are hardy little dudes. In 2012, researchers in Japan at the Kanazawa Medical University found that they can survive in a near vacuum for upwards of 30 minutes.[55]

After they crawl onto their host, they pierce with their sucking mouthparts and basically cement themselves to the animal. They can eat up to 100 times their body weight. Females will eat until they

55 Ed Yong, "Stuffed in a Vacuum and Bombarded by Electrons, a Tick Waves Hello," *Discovery*, March 15, 2012, https://www.discovermagazine.com/planet-earth/stuffed-in-a-vacuum-and-bombarded-by-electrons-a-tick-waves-hello.

look like they're going to pop, whereas males will stop feeding about halfway to that point.

Ticks are either three-host, two-host, or one-host feeders. Three-hosts means they feed and fall off of their host between each life stage. Three- and two-host feeders are the most common. One-host feeders spend their entire lives on a single animal. On average, ticks spend around ninety percent of their lives *off* of their host, even though we only notice them when they're *on* us.

In order to grow and reproduce, most species need blood. Females can lay up to 5,000 eggs at a time. Once the eggs hatch, the new babies (and the parents who still need to feed) use their "ambush strategy" to find another host. They climb onto vegetation with their front arms outstretched, waiting for some animal to pass by. They hook on when something brushes against it.

As with all arthropod orders, some tick species do not fit the normal tick description. *Otobius megnini* is a unique species. They only have a single nymphal stage (most have three). Neither the males nor the females feed, and the females lay eggs without having had a blood meal. Basically the opposite of every other tick.

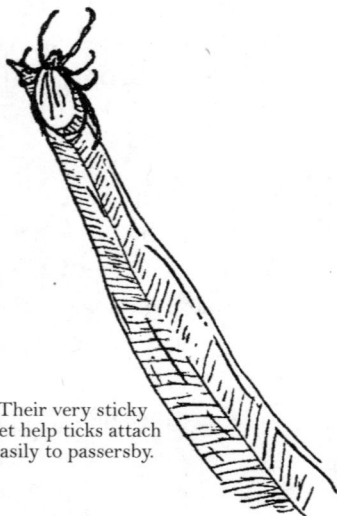

Their very sticky feet help ticks attach easily to passersby.

How Humans Can Interact with Ticks

While having a tick embedded in your skin is an uncomfortable experience, the biggest worry about a bite is potential diseases. Surprisingly, ticks cause more damage to cattle than humans. Worldwide, an estimated $7 billion-plus in tick-related losses hits the cattle industry.

Ticks transmit a greater variety of disease-causing pathogenic agents than any group of arthropods, including protozoan, viral, bacterial, and even fungal pathogens. In regards to humans, ticks carry fourteen known spreadable diseases (as of 2009). Rarely do they cause death but they can cause chronic issues, like Lyme disease. Removing them completely from your body is important for reducing chances of infection (eradicating them in the world would not be possible or helpful).

You may have heard the saying, "Tweezers are tick squeezers." If you squeeze a tick while they are attached, they regurgitate all of the blood in their body back into you, increasing your chances of disease.

However, there are a few myths surrounding the best ways to remove ticks. Similar to chiggers, people will sometimes paint them with nail-polish to encourage them to back out. You want to remove them as soon as possible, not wait for them to leave. The CDC recommends these following steps:

- Use clean, fine-tipped tweezers to grasp the tick as close to the skin's surface as possible.
- Pull upward with steady, even pressure. Don't twist or jerk the tick; this can cause the mouth-parts to break off and remain in the skin. If this happens, remove the mouth-parts with tweezers. If you cannot remove the mouth easily with tweezers, leave it alone and let the skin heal.
- After removing the tick, thoroughly clean the bite area and your hands with rubbing alcohol or soap and water.
- Never crush a tick with your fingers. Dispose of a live tick by . . .
    - Putting it in alcohol,
    - Placing it in a sealed bag/container,
    - Wrapping it tightly in tape, or
    - Flushing it down the toilet.

If you are worried about the bite, because there is a red circle around it or you are curious, bring the tick with you (in tape) to the doctor's office, along with information about where you got it, which will help determine what tests to run. Often, you can be bit by a tick with no issue. Still, it's good to know how to properly remove them and what to tell your doctor.

How Human Culture Is Shaped by Ticks

Ever heard of an Alpha-gal allergy? It's a strange, fairly new transmittable allergy from being bitten by a native US tick. First discovered in 2009, the allergy found its root in the Lone Star tick—making people allergic to red meat. Some people even found that eating red meat can send them into anaphylactic shock, sometimes even hours after consuming the meat. This allergy is most common in Texas and Florida. (Perhaps it's Mother Nature's way of fighting back against the meat industry?)

Despite these negative impacts, ticks can be useful in unexpected ways. Many of the compounds and toxins found in ticks are used in medicines for various ailments. They are being studied to see how effective they would be in heart medicine, antibiotics, stopping necrosis, and reducing tremors. One of the most impressive possible uses of tick toxin is to cure certain cancers.[56] Who knows? In a decade you may have ticks to thank for curing cancer.

Outside of the medical world, ticks don't show up much in popular culture, apart from one character. The Tick is a well-known, if not silly, comic book character who has had a long run of comics (still going), his own animated show, and two live-action shows. The Tick has even less in common with a tick than Spider-Man with a spider. He truly shares nothing but a name with the animal. Still, he is a beloved character who gives a positive portrayal of a creature with that name.[57]

---

56 Péter Apari and Gábor Földvári, "Harm or Protection? The Adaptive Function of Tick Toxins," *Evolutionary Applications* 14, no. 2 (2020), https://doi.org/10.1111/eva.13123; Jindrich Chmelar, Jan Kotál, Anna Kovaríková, and Michail Kotsyfakis, "The Use of Tick Salivary Proteins as Novel Therapeutics," *Frontiers in Physiology* 10, no. 812 (2019), https://doi.org/10.3389/fphys.2019.00812.

57 Kind of positive. The Tick is often portrayed as clinically insane but still meaning well. If nothing else, this is a more positive portrayal than ticks often get.

Quick Facts: Ticks

- These creatures don't actually have a head but instead only have mouthparts. Their brain is further into their abdomen.
- They basically cement themselves to their host ,which is what makes them so difficult to remove.
- Most of their life, around ninety percent, is spent off of their host, so we only notice them during that other ten percent of time.
- They find a host by waiting at the end of a blade of grass or vegetation with their front legs out, in the hopes of hooking onto a warm-blooded creature. Little weirdos.
- The Tick is a famous superhero character of sorts. He shares less in common with a tick than Spider-Man does with a spider.

Mites

Entomologists estimate that only about five to ten percent of all mite species have been discovered, due to their small size. Mites are found everywhere, from tropical rainforests to rocky outcrops in Antarctica, from desert habitats to the deep ocean trenches. They can be found several meters deep or even in groundwater. They occur in all types of aquatic habitats, including freshwater lakes, streams, and even hot springs.

Their role in these environments is essential, as they cycle important nutrients like calcium. They also help disperse bacteria and fungus needed for organic decomposition. In a ten by ten inch space of the forest floor, there are likely over 250,000 mites. The old European belief that they sprung spontaneously from the ground is somewhat logical since they are often found there. They act as cleaning service and pest control. All of this while often being smaller than the dot of an *i*.

How Mites Shape Our World

Some mites are more conspicuous than others like the true velvet mite, which is a bright red. Many are more subtle colors.

The fastest animal on the planet is a mite, *Paratarsotomus macropalpis*, who can run faster than any other animal. Their speed would be the equivalent of a person running 1,300 miles per hour.[58] Another mite can lift 1,180 times its body weight, which would be like a 150-pound person lifting 177,000 pounds.[59]

In order to live in all the places they do, mites needed some creative ways to spread since they have no wings and very tiny legs. They use wind, silk ballooning, hosts, and hitching a ride with others (scientific term for that is *phoresis*) as their ways of transportation to new habitats.

Some mites can cause mange in animals, both domestic and wild. Varro mites can be found on bees and in beehives. They live on honeybees to weaken them, and they can be one factor in colony collapse (with a mixture of other things like pesticides). Having mites in a hive doesn't immediately lead to colony collapse, but it is something that entomologists are trying to learn more about.

But not all mite/bee relationships are bad. Some species of Old World carpenter bees may carry several species of mites on their bodies. The female bee carries mites that scrape the cuticle of the bee larvae and remove potentially pathogenic microorganisms and fungal spores. Some mites are mutualists in the nests of sweat bees that wander over the nectar and pollen, straining potentially harmful microorganisms.

58 Federation of American Societies for Experimental Biology (FASEB). "Mite Sets New Record as World's Fastest Land Animal," *ScienceDaily*, April 27, 2014, www.sciencedaily. com/releases/2014/04/140427191124.htm.

59 Michael Heethoff and Lars Koerner, "Small but Powerful: The Oribatid Mite Archegozetes Iongisetosus Aoki (Acari, Oribatida) Produces Disproportionately High Forces," *Journal of Experimental Biology* 210, no. 17 (2007), https://doi: 10.1242/jeb.008276.

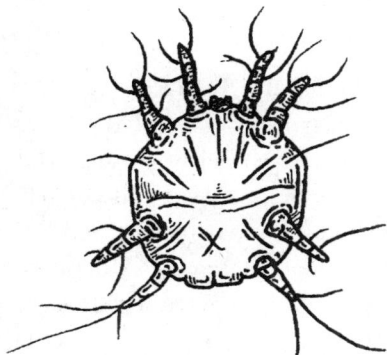

Close up of varro mite (right). Bees can become overwhelmed by these mites (left).

How Humans Can Interact with Mites

Humans are covered in mites, as are all our belongings. A mite species lives in the hair follicles of our faces and occasionally other body regions, and another lives in the sebaceous, or sweat, glands in the skin. Most people have these mites with no negative effects. Mites typically inhabit beds, chairs, and carpets in houses, and their shed skins and feces provide the bulk of the allergens in house dust.

Vacuuming regularly helps with these allergens. Some mites that hang out on humans are more noticeable and annoying, like scabies. Mites that cause scabies (a different species from body lice) are normally not a problem for healthy humans. They can, however, cause serious issues for adults with health problems, leading to crusted scabies. This advanced form of scabies infestation is much more difficult to eradicate and makes the person more contagious since there can be upwards of two million mites on a single person.

Another annoying species is the chigger mite. Chiggers are a parasitic mite found in the woods. The larvae attach to the host, spit out some digestive enzymes and then suck up the skin cells. They don't lay eggs in you, they don't burrow, and they don't drink blood, but they do cause welts that are unbearably itchy. By the time you start itching though, they are already gone.

How Human Culture Is Shaped by Mites

Numerous mites are important to certain types of cheese. Yes, there are cheese mites. Casu martzu is a rotting specialty cheese with maggots and mites. There is a similar cheese in Egypt called mish. Some popular French cheeses, like Mimolette, have mites actively eating the rind. The US didn't allow the importation of Mimolette until recently because of the living beings on it. Now, you can buy it at most stores that sell specialty cheese. It is a yummy orange cheese with a softer rind (thanks to the mites).

Quick Facts: Mites

- They are super tiny but appear in soil in great numbers: 250,000 for every 10x10 piece of land.
- They are essential to soil health and decomposition, as they help aerate the land and break down detritus.
- All humans have thousands of mites living on them at any given time. They don't harm us and help to clear dead skin. Some people have sensitivities to mites, so vacuuming regularly helps with that.
- They help make some delicious cheese like Mimolette, a French cheese with its rind softened by mites.
- The fastest animal on the planet is a mite, which can run faster than any other animal. Their speed would be the equivalent of a person running 1,300 miles per hour.[60] Another mite can lift 1,180 times its body weight, which would be like a 150-pound person lifting 177,000 pounds.[61]

60 FASEB, "Mite Sets New Record as World's Fastest Land Animal," www.sciencedaily.com/releases/2014/04/140427191124.htm.

61 Heethoff and Koerner, "Small but Powerful," https://doi: 10.1242/jeb.008276.

CHAPTER 16: MILLIPEDES (CLASS DIPLOPODA) AND CENTIPEDES (CLASS CHILOPODA)

Similar to ticks and mites, millipedes and centipedes share similarities but are drastically different. They both have long bodies and many legs. One is aggressive, one is not. One is predatory, one is a detritivore. Both are good for your garden for different reasons, and they don't often want to be indoors (apart from the house centipede).

Once you know what to look for, it is easy to tell the difference between these two.

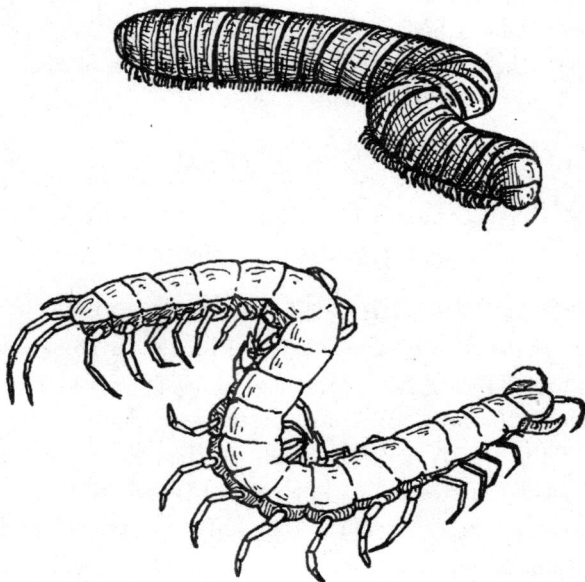

Millipede (top) and centipede (bottom)

	Millipedes	**Centipedes**
Legs	Two pairs per segment, on the underside of the body	One pair per segment, out to the side of the body
Movement	Slow	Fast
Feeding	Most are detritivores, some herbivores, no venom	Most are carnivores with front legs modified into venomous fangs
Body Shape	Rounded	Flat

Millipedes (Class Diplopoda)

While *millipede* means "thousand legs," scientists didn't find a species with that number of legs until 2020.

The largest millipedes today are around 15 inches (the Giant African Millipede), which only have about 400 legs. When these guys showed up on the scene 428 million years ago, some of their ancestors were up to 6 feet long, 2 feet wide, and as tall as a large dog.

People keep Giant African Millipedes as pets, but can you imagine having a 6 foot long millipede instead? I'd ride it everywhere and it could eat all of my compost. Like a long, many legged, rideable, trash-eating cow.

How Millipedes Shape Our World

Around 12,000 species of these harmless little (or sometimes not so little) detritivores have been identified. They have two pairs of legs per segment, but this can be difficult to see since the legs are held underneath their body. Even with all of these legs, they are pretty slow and most likely to curl up in a spiral to avoid being messed with. Some secrete a smelly chemical that only has an effect on smaller animals, not humans.

Because millipedes don't move very fast and are fairly small, many of their ranges are restricted. In contrast, to other animals who have hosts or silk to balloon them away, millipedes have to rely on crawling to new places.

They can live up to ten years, but they more commonly have a lifespan of a year or two. As they grow, they continually molt. Each molt, they add further segments and legs, a mode of development known as anamorphosis. Some species are parthenogenetic, like many of their arthropod kin, so there may be few or no males of the species.

Their presence in ecosystems is important and they have a role similar to mites and other decomposers. They facilitate microbial decomposition of the leaf litter.

Many live on their own. Some have been known to form large clusters to avoid being eaten. Some live in ant colonies mutually; some have relationships with mites that they help disperse to new places; some have relationships with mosses that grow on the bodies of the millipedes and help to camouflage them.[62]

They have many natural predators, including a family of assassin bugs that only eat millipedes. Apart from their defenses of rolling into a ball or secreting a foul liquid, some types of millipedes have developed bristles. These bristly millipedes use their hair—similar to the hair on a tarantula's abdomen—to ward off small insect predators like ants.

How Humans Can Interact with Millipedes

With all of the arthropod species, people are continuing to discover new species. One millipede species was found in Los Angeles, California in 2023.[63] It has 486 legs and is almost translucent. It was found by

62 Shirley Daniella Martínez-Torres, Álvaro Eduardo Flórez Daza, Edgar Leonardo Linares-Castillo, "Meeting between Kingdoms: Discovery of a Close Association between Diplopoda and Bryophyta in a Transitional Andean-Pacific Forest in Colombia," *International Journal of Myriapodology* 6 (2011), https://doi.org/10.3897/ijm.6.2187.

63 Margaret Osborne, "New 486-Legged Millipede Species Found Near Los Angeles," *Smithsonian Magazine*, July 28, 2023, https://www.smithsonianmag.com/smart-news/new-486-legged-millipede-species-found-near-los-angeles-180982623/.

two citizen scientists who posted it on iNaturalist. Again, taking pictures for that app could result in a new species being identified. Anyone can post, so the next time you see something interesting (or even mundane), snap a photo to post there.

You may find millipedes in your plants outside. If you find one inside, be sure to relocate it. It needs soil and decomposing matter to live. It does not want to be in your house. Remember, they don't bite and often curl up when touched so they are easy to transport back into your yard or some nearby grass.

How Human Culture Is Shaped by Millipedes

In *A Bug's Life*, P. T. Flea who runs the circus uses millipedes to pull his carriage, which seems extremely inefficient since they are so slow. *A Bug's Life* is a great, positive representation of bugs. And now Disney has a program called *A Real Bug's Life* about insects in the real world. No, I have not yet watched it (at least by the time I'm typing this) and no, this is most definitely not a paid advertisement for Disney. Or is it? (No, it's definitely not.)

Quick Facts: Millipedes

- They don't bite and are harmless detritivores.
- They are important to soil health and decomposition in forests and other wooded areas.
- They are rounded, with their legs appearing under the body.
- Four hundred million years ago, they could be six feet long and two feet wide. Now, the biggest one is fifteen inches long.
- Each time they molt, they add a new section and a new pair of legs. Even with the name millipede, most do not have a thousand legs.

Centipedes (Class Chilopoda)

The evil twin (not really) of the millipede is the centipede. Centipedes would rather not deal with you, but it is good to know about them, outside of the myths and misinformation. In the US, most centipedes are small and the species most people come into contact with is the House Centipede. I love bugs, but seeing one of them run up the outside of my shower curtain while I was showering was something I will never forget. I've never screamed so loud. Those guys move *fast*, which makes sense given the layout of their legs.

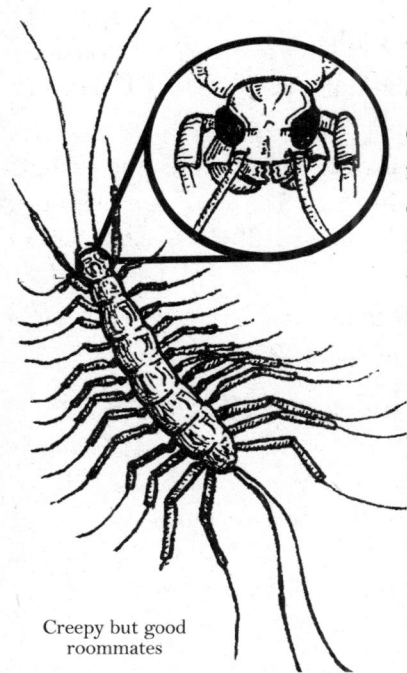

Creepy but good roommates

How Centipedes Shape Our World

Of the three thousand identified species, most centipedes are carnivorous predators. Speed is important to centipedes on the offensive and defensive side. Not all species chase their prey.

The biggest centipede in the world is twelve inches long, the Amazonian Giant. This species hangs from the roof of caves and waits for a bat to fly by. It then catches and eats the bat.

The majority of centipedes don't eat bats but have been known to eat lizards, snakes, small mammals, and mostly other arthropods. Their bite is venomous, but, unlike scorpions that have a stinger, or spiders that have fangs, centipedes have modified front legs that are venomous, which are used in a fang-like way.

They also have back legs that stick out and aren't used for walking. These pairs of legs are called "ultimate legs" and may aid in reproduction or can be rubbed together to make noise that scares predators. Those defenses don't always work and so centipedes are commonly eaten by larger animals like mammals or coyotes. Animals that have centipedes on the menu will often eat the head first to avoid any potential bites.

How Humans Can Interact with Centipedes

Centipedes have been known to bite humans, often when the human is unaware that the centipede is there and goes to sit or step on it. The bite can be painful but is almost never fatal, especially in the US. The species that cause the most pain are very large and are not found in the States. Since they are most active at night, people rarely have issues with them.

I found one at the bottom of a pool once in Arizona (the US does have a somewhat large desert centipede). I had never seen anything like it, as midwestern US versions are much smaller. Its modified front legs were impressive.

Humans also eat centipedes. In Asian markets, centipedes are a common sight, on a stick like a kebab, sold by street vendors.

How Human Culture Is Shaped by Centipedes

While some humans enjoy eating centipedes, others are completely terrified of them. Movies have taken advantage of that and created some centipede-related horror films. Two horror movies have been released with the title *Centipede* (1994, 2004). And who can forget the trilogy of *Human Centipede* (2016) movies? (Hopefully, most people have forgotten.) '

In a more useful (and positive) capacity, researchers at Osaka University created a centipede robot for search and rescue, as these creatures can go into small cracks and wind in various directions.[64] Their segments allow for a wide range of motion which could be useful in rubble.

---

64 University of Osaka, "Robot Centipedes Go for a Walk," EurekAlert!, May 29, 2009, https://www.eurekalert.org/news-releases/990577.

QUICK FACTS: CENTIPEDES

- Unlike millipedes, centipedes have a modified pair of forelegs that contain venom. They bite only when threatened.
- There is a giant centipede who waits on the ceiling of caves and catches bats that fly by, which is ... . something else.
- Centipede females can be caring mothers who protect and feed their newborns until they are ready to go out on their own.
- Many animals eat centipedes and start by eating the head so as not to be bitten.
- Researchers are trying to create robots that move in a similar fashion to centipedes, as they can easily make their way through small cracks and traverse treacherous debris. The robots could be very useful for search and rescue operations.

CHAPTER 17: ISOPODS (ORDER ISOPODA)

R ollie pollies, pill bugs, woodlice—whatever you call them,[65] these little cuties are one of the few bugs that most people like. They were one of my favorites as a kid, and I would walk around with pockets full of them, to pull out and show people throughout the day. (It probably wasn't the best idea and a lot probably died, but I was clearly very enthusiastic about them. RIP to all the rollies I killed on accident.) To this day, I still get excited to see them. The other day I was walking into work, I saw one and in a high-pitched (like "look a puppy") voice said, "Oh, rollie pollie!" It made my day.

Rollie pollies are actually not even arthropods, but crustaceans. That means that they are more closely related to a crab than a spider. They have been around for about three hundred million years.

The name *isopod* comes from Greek, meaning "equal foot," since they generally have one pair of legs per segment.

Many isopods are aquatic and live in both fresh and salt water environments. The ones that are terrestrial evolved from those who are in the sea, but terrestrials no longer have the capability to be in water.

Scientists have only officially identified around 10,000 species, but they estimate that there are upwards of 123,000 species total. The majority of the unknown species are most likely aquatic.

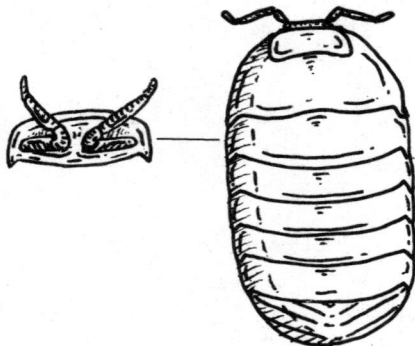

---

65 Some other names include "armadillo bug," "chisel pig," "potato bug," and "woodbunter."

How Isopods Shape Our World

Isopods love eating decomposing matter, which makes their place in forest ecosystems important. Also, other animals eat them, making their part in the food web essential as well.

Similar to millipedes, pill bugs use the same defense mechanism of curling into a ball. There is even a millipede that looks very similar to a pill bug, called the pill millipede. Their similar appearance is an example of convergent evolution.

A new giant isopod was discovered in 2022. It lives about eight hundred meters underwater and is about ten inches long.[66] In the same year, researchers discovered that an aquatic isopod pollinates a type of seaweed. The process looks a bit different with underwater pollination, but the isopods carry sperm-like bodies from male to female plants. The scientists who discovered this wonder if this evidence supports the idea that pollination actually started in the ocean before it started on land.[67]

Terrestrial rollie pollies may be related to aquatic species, but they are not able to withstand water. They need to stay moist but can't survive in water itself. Somewhere on the evolutionary tree, they split and became exclusively terrestrial. If you ever see one in a puddle or a pool, know that they can't stay there too long without dying. They drown. Their little lungs are in their butt, so you may see them airing out their derriere in an attempt to remove water from their bodies.

How Humans Can Interact with Isopods

You may see isopods in your house, but chances are they are there because you have a leak somewhere. They do not spread disease, so if you do see some, know that you and your family will be okay.

66 Taylor and Francis Group, "New Giant Deep-Sea Isopod Discovered in the Gulf of Mexico," Lab Manager, August 10, 2022, https://www.labmanager.com/new-giant-deep-sea-isopod-discovered-in-the-gulf-of-mexico-28602.

67 Carissa Wong, "Small Woodlouse-Like Crustacean 'Pollinates' Red Seaweed with Its Legs," *New Scientist*, July 28, 2022, https://www.newscientist.com/article/2331419-small-woodlouse-like-crustacean-pollinates-red-seaweed-with-its-legs/.

Some people want them in their house, but under their rules. Isopod enclosures are a way to keep them as pets but also have some greenery that you don't have to take too much care of. Bioactive terrariums include both plants and isopods, which then reduces the amount of cleaning the owner has to do of the enclosure.

Rollie pollies are known to roll up into a ball as a defense mechanism, but researchers have determined that it also helps them conserve water.[68] I'm sure you've encountered one that rolled up when you touched it.

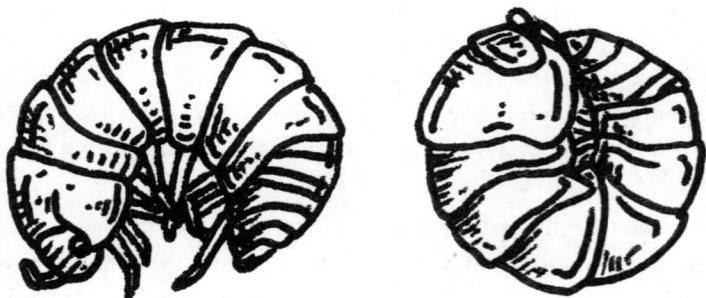

Multi-purpose rolling

Living at depths only accessible to humans via submarine, aquatic isopod species are less likely to be encountered by people. They can be much larger than terrestrial species. The Giant Isopod (*Bathynomus giganteus*) can be close to twelve inches long.

Even if a human were to interact with them, aquatic isopods are very gentle and only eat decaying animals, like crabs and other dead ocean creatures. They can go for long periods of time without eating (think five years) and can be seen in some aquariums or invertebrate

68 Joshua Cassidy, "Pill bugs Emerged from the Sea to Conquer the Earth," *PBS News Hour*, February 9, 2017, https://www.pbs.org/newshour/science/pill-bugs-emerged-sea-conquer-earth.

zoos around the world. There still isn't much known about them and their behaviours in the wild.

How Human Culture Is Shaped by Isopods

Scientists in Japan have been trying to determine possible uses of pillbugs. In one study, a small 3D-printed arm held the pill bugs as they were lowered to different types of material. They were able to lift a decent amount of weight. Scientists have no real idea what to do with the information or how it could change robotics, but they did the study anyway. Now they have the information should they ever need it.[69]

Pill bugs don't show up in culture as much as one might imagine, but they are small and innocuous. There are two pill bugs in *A Bug's Life,* Tuck and Roll, who don't speak the same language as anyone else in the circus so they always seem to be out of the loop and doing their own thing. They are comedic relief and get to show off their rolling skills. And researchers have determined that rollie pollies may have personalities, so that's cool and adorable.[70]

They may be a good gateway species. A gateway species is an animal that introduces people to a family of animals (like invertebrates) to show people that they are not scary. Some people also call butterflies and dragonflies gateway species.

People do find pill bugs cute, and they can often be found as earrings, necklaces, or tattoos (I'm planning to get one on my kneecap).

The tongue louse is the subject of a movie that exaggerates the abilities of an isopod that takes over the tongues of fish. Less cute,

69 Alex Wilkins, "Scientists Want to Give Robots Hands Made from Living Woodlice," *New Scientist,* June 20, 2023, https://www.newscientist.com/article/2378715-scientists-want-to-give-robots-hands-made-from-living-woodlice/.

70 Gergely Horváth, László Zsolt Garamszegi, Judit Bereczki, Tamás János Urszán, Gergely Balázs, and Gábor Herczeg, "Roll with the Fear: Environment and State Dependence of Pill Bug (Armadillidium vulgare) Personalities," *The Science of Nature* 106 (2019), https://doi: 10.1007/s00114-019-1602-4.

more terrifying. But to be clear, these creatures are not capable of taking over human tongues.

Many species are still exclusively aquatic so they have adapted in some ways that terrestrial ones have not. The most intriguing feeding specialization that an aquatic isopod has belongs to the tongue-eating louse (*Cymothoa exigua*, not a louse but an isopod). The bug takes the place of a fish's tongue by latching on and causing the fish's tongue to eventually fall off. It essentially takes the place of the tongue and then eats some of the food the fish takes in. The fish continue to live a normal life, but evidence points to affected fish having less robust health, like being underweight or having less hardy young.

Quick Facts: Isopods

- There are both terrestrial and aquatic isopods. Most terrestrial isopods are unable to take overly wet conditions and end up drowning.
- They curl up when threatened but also when they want to conserve water.
- They eat a lot of detritus and are important for soil health, decomposition, and the food web.
- Researchers have found that rollie pollies have distinctive personalities.
- Recently scientists found that an aquatic isopod is responsible for "pollinating" some underwater plants. The pollination underwater looks a bit different but appears to be important to the life of the plants. Terrestrial isopods are not associated with pollination.

Chapter 18: Aquatic Inverts—Crayfish, Bivalves, and Horseshoe Crabs

There are serious numbers of aquatic invertebrates (including the isopods and mites in previous chapters). This chapter is kind of a catchall for some less discussed, but super important bugs, two which are arthropods and one which is a mollusk.

Aquatic invertebrates are crucial to healthy water systems, which humans should definitely care about, since we rely on that to live.

The Missouri Department of Conservation created a lovely list as to why aquatic invertebrates are so important:

- They are essential to the food chain in fresh waterways for fish, reptiles, mammals, and birds.
- They are indicators of water quality: they do not tolerate pollution, so it is important to watch fluctuations in their populations.
- They eat pest species such as mosquitoes.

Let's take a look at three of these creatures to see why they are key members of healthy aquatic environments.

Crayfish (Order Decapoda)

These inverts look like small lobsters, which makes sense considering they are very closely related; crayfish are crustaceans. Both are part of the Decapoda order (Greek for "ten-toed"). Unlike lobsters, crayfish are found exclusively in freshwater habitats. They are also quite a bit smaller usually, but some people locally call them "lobsters." The word *crayfish* comes from Americanizing the French word for them, *crevice*.

Different parts of the US have different names for them: crawdads, crawfish, lobsters, mudbugs. In Missouri, I grew up calling them "crawdads" or "crawfish." Australians apparently call them "yabbies," which is fun.

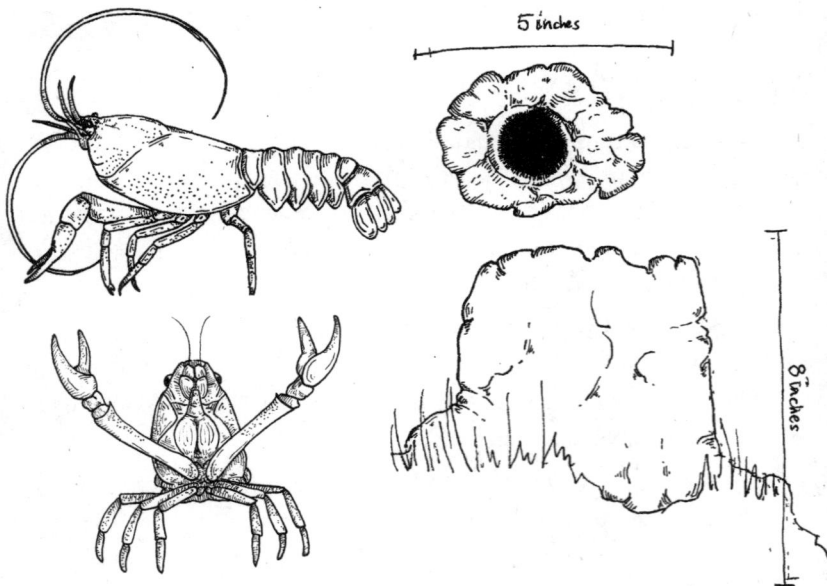

Some people find crayfish mud tubes annoying, as there can be several that appear in one yard, similar to mole damage.

How Crayfish Shape Our World

Like their arachnid relatives, crayfish have a cephalothorax and an abdomen. Also like arachnids, they have a similar defensive posture: front two legs out and raised. Additionally, they have the ability to send themselves hurling backwards in the water with the force of their tail, moving away from the predator.

Their walking legs have tiny little claws at the end of each one, in addition to their large claws. Their claws aren't used much for food, as they are scavengers. As opportunists, they've been known to eat a lot of different things, including dead animals, their own poop (coprophagy), and their molted shell.

Eating all of those things has helped them be successful over the last 300 million years. In North America alone, there are 330 species. Some of these species have been introduced. The American crayfish was introduced to Europe and is now considered a pest to native frogs.

Similarly, a species called the rusty crawfish was found in Colorado recently. Interestingly, these crayfish are native to the US but are mostly found in the Ohio river. When they have been spotted in other bodies of water across the US, they cause issues with native crayfish species of that area.

Wildlife biologists are on high alert to make sure rusty crawfish numbers don't spread. Robert Walters, a Colorado Parks and Wildlife Invasive Species Program worker, stated, "They themselves are a very aggressive crayfish species, which leads to them outcompeting our native species for resources and also makes it so that they themselves are not a good prey species for our sportfish."[71]

How Humans Can Interact with Crayfish

Spreading of non-native species is common. In almost every situation, humans are the cause. They can also be the cause of spreading native species to new locations, which is what people think happened with the rusty crawfish. Some people use them as bait while fishing which may explain how they've spread outside of the Ohio river. Some people keep crayfish as pets and biologists worry that non-native species may be spread by release of these pets into waterways.

Living around a year or more, some crayfish normally dwell in streams and hide under rocks for cover. These species are ones most often kept as pets. Other species create distinctive mud tubes up to eight inches high. Homeowners may find them annoying, as the tubes can be tripping hazards.

In 2023, a new species was found in Tennessee by two grad students. They were there looking for frogs but saw this vibrant

71 Alex Hager, "Officials Are on Alert after the Rusty Crawfish Shows Up Near the Colorado River," KEUR, September 27, 2023, https://www.kuer.org/2023-09-24/this-pesky-crayfish-showed-up-in-colorado-and-wildlife-managers-are-on-high-alert.

blue crawfish. They wondered at it for a moment, then got back to what they were there to look for initially. Later that night, one texted the other to ask if they had kept the specimen. They both regretted not capturing the bright blue boy, so they went back the next day. They searched for hours and finally found it. They were glad they did because soon after, it was identified as a previously unknown species. Again, a great reminder that there are always new species to discover in your own backyard.

How Human Culture Is Shaped by Crayfish

In the US, crawfish are extremely important to Cajun culture. They are the official state crustacean of Louisiana, and adorn all types of gear: jewelry, t-shirts, stickers, plaques. Large crawfish boils are common for large parties during crawfish season. A boil includes throwing whole crawfish, potatoes, onions, corn, and spices into a ginormous pot to cook everything together. The pots are large enough to fit crawfish needed to feed hundreds of people.

An important note: Research shows that crayfish do not die immediately when boiled alive, similar to lobsters. They respond to pain in a similar way to mammals. Some places will kill them before boiling. Perhaps something to keep in mind should you need to boil any.

Quick Facts: Crayfish

- There are many names for crayfish depending on where you're from: crayfish, crawfish, mud lobsters, river lobsters, and yabbies.
- Even species native to the US can be invasive in other parts of the country, like the rusty crawfish. It is native to the east coast and the Ohio River, but it has been found in Colorado. Parks and Rec is concerned about it outcompeting native Colorado crayfish.
- They are an important part of the food web for vertebrates like raccoons and weasels.

- They are also an important food and cultural symbol in Louisiana. Crawfish boils and crawfish on clothing are all common in that part of the US.
- Almost all crayfish live in freshwater, with very few species living in brackish or salt water.

FRESHWATER BIVALVES (CLASS BIVALVIA)

Freshwater mussels, also called freshwater bivalves (meaning "two shells"), are really the unsung heroes of rivers. They are mollusks (phylum Mollusca, class Bivalvia), meaning that they are not an arthropod. They are, however, an unbelievably important invertebrate in water systems.

Many people know about mussels, mostly as a food, or have heard stories of invasive species, like the Zebra mussel. It turns out their lives are much more interesting than often portrayed. The Xerces Society educates and advocates for mussels and mussel conservation. Their literature on mussels is extensive, so if you have any interest in these creatures, the society's website would be a great place to start.

HOW FRESHWATER BIVALVES SHAPE OUR WORLD

Mussels can live up to hundreds of years and are affected by global warming: droughts, floods, heat waves. All of these events are detrimental to their populations. More than seventy percent of all species of North American freshwater mussels are considered endangered, threatened, or of special concern.[72] According to the Xerces society, "in 2021, two dozen plant and animal species were removed from the endangered species list. It was nothing to celebrate however; seven freshwater mussel species were removed because they are thought to now be extinct."[73] This news is devastating for a variety of reasons. Mussels are not only important as a food source for humans but also other animals.

72 Justine Wheeler, "Putting Muscles on Your Mind," Xerces Society, October 31, 2016, https://www.xerces.org/blog/putting-mussels-on-your-mind.
73 Emilie Blevins, "The Merit of Mussels," Xerces Society, October 14, 2021, https://www.xerces.org/blog/merit-of-mussels.

Freshwater mussels filter dangerous chemicals and heavy metals. Because of these filtering abilities, water exits mussels much cleaner than when it went in. Since mussels live for long periods of time and are sensitive to environmental changes, they are good bioindicators. They are also relatively stationary so you can track the health of a stream over time. The Xerces Society states,

> "As we can look at tree rings to measure the age of a tree and learn about the climate and conditions of a forest over many years, mussel biologists can determine the age of freshwater mussels by counting the rings etched into shells and make observations about long-term stream health."[74]

How Humans Can Interact with Freshwater Bivalves

People have eaten mussels for thousands of years, but with dwindling numbers and how much they filter impurities, it may not be the best idea in this day and age. We shouldn't even eat the invasive ones.

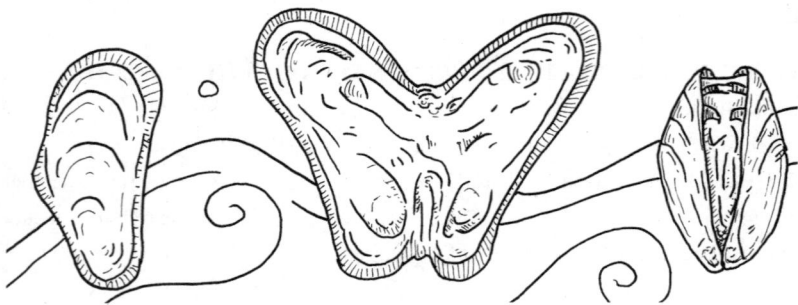

One such invasive species? Originally from Russia, zebra mussels are responsible for the near extinction of many species in the Great Lake system by outcompeting native species for food and by growing on top of and suffocating the native clams and mussels. They block pipelines. They clog water intakes, at both municipal water and

74 "About Freshwater Mussels," n. d., https://www.xerces.org/endangered-species/freshwater-mussels/about.

hydroelectric company systems. They affect all classes of algal species, resulting in a shortage of food to native freshwater mussels and fish. They are also believed to be the source of a deadly avian botulism poisoning that has killed tens of thousands of birds in the Great Lakes since the late 1990s. All of these behaviors have led to them being considered one of the most invasive aquatic species ever. All attempts to eradicate them have failed. And you can't even eat them. Since they filter water, they tend to accumulate pollutants and toxins, so experts recommend against consuming zebra mussels.

Most current news about mussels focuses exclusively on the invasive species, which is understandable; however, it is also important to discuss and teach about native ones that need our help. We need to protect our waterways, and in doing so, protect all of the important invertebrates (and other animals) who live there.

How Human Culture Is Shaped by Bivalves

In the seventeenth, eighteenth, and nineteenth centuries, shells from bivalves (both aquatic and freshwater) were used in various ways by humans. Some used small pieces of shell as currency. Some people started to make buttons out of them. At that time, the makers of the buttons didn't realize the harm they could be doing to mussel populations. People also made decorations out of them. They were fashionable in many ways.

Collecting mussels, regardless of intended use, is not encouraged since many species are endangered or threatened. Unless, of course, you know they are invasive, like the zebra mussel, and you make sure they don't end up in another body of water.

Quick Facts: Bivalves

- Most bivalves can live up to one hundred years.
- They are incredibly susceptible to water pollution and are known for their ability to act as kidneys of waterways.

- Shells were once used to make buttons and currency in the past. This led to a steep decline in numbers. There are three hundred native mussels to the US, but seventy-one percent of them are considered threatened or endangered.
- An invasive species, the zebra mussel, has wreaked havoc on native species by overcrowding them and spreading rapidly.
- Scientists discourage people from eating wild mussels since they filter harmful chemicals and metals.

Horseshoe Crabs (Family Limulidae)

I have friends who grew up or spent a lot of time on the east coast growing up. Apparently during horseshoe crab mating season, you could see hundreds of dead ones littering the beach to the delight of many birds. Apparently, it was quite the sight to behold, if not a bit confusing at first. I have never seen this but I hope to make it to the east coast for breeding season some day.

These living fossils dwell in saltwater only. They are some of the oldest living arthropods: 480 million years old. Also, unlike crayfish and freshwater bivalves, horseshoe crabs are closely related to spiders, scorpions, etc. in the subphylum Chelicerata. They also are related to other saltwater arthropods, like sea spiders and sea scorpions.

Only four species of horseshoe crab exist worldwide, with one living in the US on the East Coast, the American horseshoe crab. They are alien-looking creatures who eat worms, small fish, small crustaceans, and mollusks. Their protective shell, or carapace, is a large horseshoe shape with a thin tail. Their underside has four pairs of walking legs, and males have two claws that they use to hang on to the females while mating.

How Horseshoe Crabs Shape Our World

Their body parts can grow back or heal from excessive damage: legs and tails grow back, cracks in carapace heal. Tails, although they may look like a stinger, are not for defense but are instead to help flip them over should they get stuck on their backs.

Females are larger than males, and in the biggest species, females can get up to thirty-one inches long and weigh over nine pounds. They use their back legs to push themselves around on land and the ocean floor. When they swim, they swim upside down.

During mating, the males attach themselves to the back of the female, with their claws. As the female moves and lays eggs, the male then gets dragged over those eggs and fertilizes them. It's pretty humorous. Females can lay up to 120,000 eggs in a year.

Their eggs are an important source of food for some shore birds; so important in fact that the reduction in horseshoe crab numbers causes issues for the birds.

The decline in numbers is linked to a few things. Asia considers them a food delicacy which plays a factor in their being overharvested

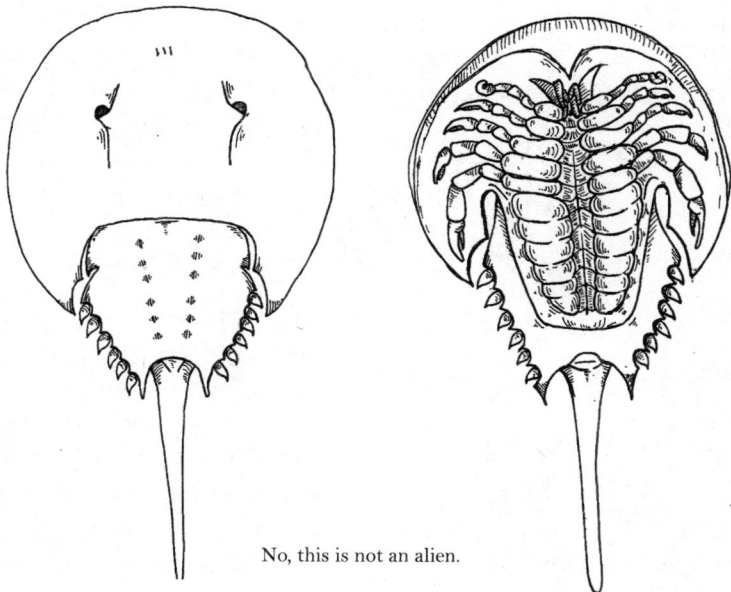

No, this is not an alien.

there. Both Asia and the US are also seeing declines in numbers because of destruction of habitat and shoreline development, use in fishing, plastic pollution, and use for scientific research and advancements.

In Taiwan and Hong Kong, the species is listed as endangered on the International Union for Conservation of Nature's (IUCN) Red List, specifically because of overexploitation and loss of critical habitat leading to a steep decline in population size.[75] Many wildlife organizations are trying to get the American horseshoe crab protected by the Endangered Species Act.

How Humans Can Interact with Horseshoe Crabs

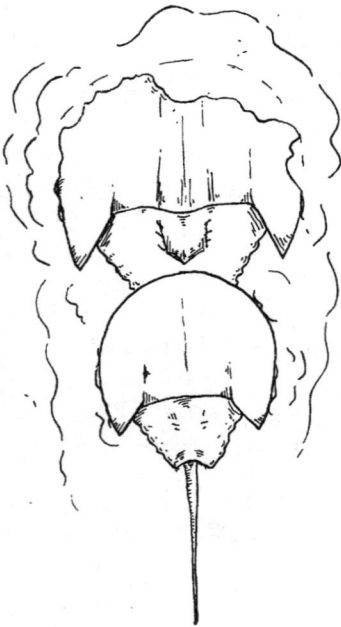

This image reeks of desperation. Anthropomorphizing animals isn't always a great idea, but I imagine the female being like, "Ugh, I keep trying to move away and he will NOT let go."

Horseshoe crabs' bright blue blood is important for vaccines, so hundreds of thousands of horseshoe crabs are milked each year for their blood. Most live through the process. If they do, it has been noted that their ability to lay eggs depletes and they aren't as strong because they've lost half of their blood. Unfortunately, captive breeding for them hasn't been very successful so scientists have to capture wild specimens.

The reason their blood is so important to vaccines is that it clots around toxins. It is used to determine if vaccines have been contaminated. A synthetic version of their blood has been created and many countries

75 Akbar John, Paul K. S. Shin, Mark L. Botton, Glenn Gauvry, S. G. Cheung, and Kevin Laurie. "Conservation of Asian Horseshoe Crabs on Spotlight," *Biodiversity and Conservation* 30, no. 1 (2020), https://doi.org/10.1007/s10531-020-02078-3.

have adopted it.[76] The US does not believe that it is as accurate as real blood, so they still only use the real thing. A continued decline in population horseshoe crab numbers might make the synthetic version the only possible option.

Currently, there are moratoriums on catching them for any reason (fishing, milking, eating) in South Carolina and Delaware. Delaware's moratorium has been put into place indefinitely. South Carolina has banned them as bait indefinitely, as well as banning the capture of them for anything else during a two month span (during their breeding window). Massachusetts is now following in their footsteps, at least during the breeding window.

How Human Culture Is Shaped by Horseshoe Crabs

Horseshoe crabs are awe-inspiring because of their strange shape, size, and presence on the beach.

In the *Fallout 4* video game, as we've seen with other arthropods, horseshoe crabs who have evolved in a nuclear soup become giant, aggressive beings. These mirelurks have many similarities to horseshoe crabs but are far bigger and more aggressive. It is still a fun inclusion of a common creature found along the east coast, even if it is not an accurate representation.

76 Katie Pavid, "Horseshoe Crab Blood: The Miracle Vaccine Ingredient That's Saved Millions of Lives," National History Museum, January 15, 2021, https://www.nhm.ac.uk/discover/horseshoe-crab-blood-miracle-vaccine-ingredient.html.

Quick Facts: Horseshoe Crabs

- Horseshoe crabs are arthropods, not crustaceans or isopods, and are some of the oldest arthropods from 480 million years ago.
- They can regrow limbs that they lose and fix cracks in their carapace (their hard outer shell).
- Their blue blood is extremely important and helpful to humans: we use their blood to test whether or not vaccines are contaminated. A synthetic version is being perfected so that the crabs will no longer need to be caught and milked.
- There are only four horseshoe crab species in the world and all are threatened or endangered due to human factors.
- Their representation in *Fallout 4* is a bit over the top. Mirelurks are gigantic, fast-moving, and aggressive, whereas horseshoe crabs are shy, slow-moving, and do not want anything to do with humans.

CHAPTER 19: SNAILS AND SLUGS (CLASS GASTROPODA)

Let's talk about gastropods—otherwise known as snails and slugs (all of which are mollusks, part of phylum Mollusca). Their Latin name means "stomach foot" in Greek, basically because they used their belly as a foot.

The oldest known fossil in this class is from 500 million years ago. Over 85,000 known species have been identified and only 2,400 of those species are terrestrial.

Determining where gastropods fall on the evolutionary tree has been a topic of debate for a century. What we all agree on is that they are more closely related to octopuses than they are insects.

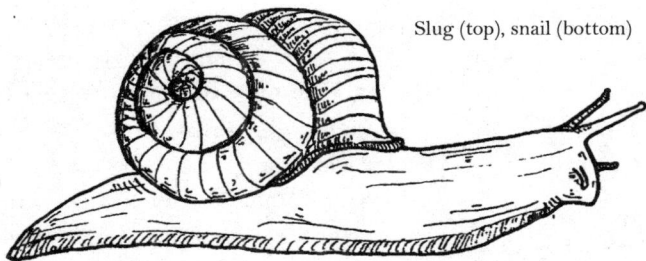

Slug (top), snail (bottom)

Biologists estimate that there are most likely upwards of 200,000 species, meaning that there are still over 100,000 species yet to be identified. Since the 1500s, 444 species have gone extinct and there are

18 species that are extinct in the wild but exist in captivity.[77] Species can be found everywhere from deep volcanic trenches to completely dry deserts.

So acrobatic.

Initially, all gastropods were aquatic and then some moved onto land. Terrestrial slugs and snails can survive a brief time in water but will drown if they are there for too long, similar to terrestrial isopods.

How Snails and Slugs Shape Our World

Most have four tentacle-like things that come out the end of their body. Those tentacles have eye spots that they use for very basic sight. Their most useful and sophisticated sense is tied to the olfactory gland that is found closer to the base of their tentacles.

Snails have a one-piece shell and slugs have no shell (but some do have an internal one). They both have a hard, spiky tongue of sorts (called a "radula") that they use to scrape off food from rocks and whatnot. If you've ever held one, you may have felt them scraping on your skin. It doesn't hurt but does feel strange.

They live for two to three years on average, but some can live upwards of ten years.

77 Claire Régnier, Benoît Fontaine, and Philippe Bouchet, "Not Knowing, Not Recording, Not Listing: Numerous Unnoticed Mollusk Extinctions," *Conservation Biology* 23, no. 5 (2009), https://doi:10.1111/j.1523-1739.2009.01245/.

Many species can self-fertilize and have both male and female sexual organs.

Slugs mating can be a crazy sight. They look like their insides have been smashed out of their body and they are dripping down (but it's actually giant penises you see). Mating can take between two and twelve hours, with the females laying about one hundred eggs afterwards.

Some species shoot "love darts" at potential mates. Research shows that the darts may increase the reproductive fitness of the successful shooter by promoting the survival and utilization of its sperm.[78] Some scholars also think it is where the idea of Cupid's arrows stemmed from.

How Humans Can Interact with Snails and Slugs

In 2024, nine new snail species were found in Papua New Guinea.[79] All of the species live in a biodiversity hotspot. Areas like that are often at higher risk of losing species and the space where the new species were discovered is no different. *National Geographic* describes hotspots best:

> "Biodiversity hotspots make up about 2.3 percent of Earth's land surface, but 44 percent of the world's plants and 35 percent of land vertebrates live in these regions. Most plants in a biodiversity hotspot are endemic, meaning they are not found anywhere else on Earth."[80]

They go on to say that hotspots are currently in crisis, meaning that they have lost more than 70 percent of their wildlife because of human factors.

78 Ronald Chase, "The Function of Dart Shooting in Helicid Snails," American Malacological Bulletin 23, no. 1 (2007), https://doi.org/10.4003/0740-2783-23.1.183.

79 Jiayu Liang, "Nine New Snail Species Discovered in Papua New Guinea, A Biodiversity Hot Spot at Risk," Florida Museum, January 16, 2024, https://www.floridamuseum.ufl.edu/science/nine-new-snail-species-papua-new-guinea/.

80 "Biodiversity Hotspots," *National Geographic*, n. d., https://education.nationalgeographic.org/resource/biodiversity-hotspots/.

If biodiversity hotspots contain 35 percent of all vertebrates, imagine how many endemic invertebrate species, like slugs and snails, they must contain.

Some people like slugs and snails, until they decide to eat out of their garden. The UK no longer qualifies them as pests because of their importance as decomposers. They can also be another important macroinvertebrate in freshwater streams. Their presence also helps to determine overall water health. Many terrestrial and freshwater species are threatened for the same reasons as other species: loss of habitat, pollution, and climate change.

They are known to carry parasites, which often aren't an issue for humans. Sometimes, strange things happen though. Have you ever been dared to eat an earthworm, frog, slug, or snail? Unfortunately, that can lead to some serious consequences. Both slugs and snails are known to carry rat lungworm, which prefers rats but can be deadly to humans.

A poor nineteen-year-old in Australia was dared to eat a snail by his friends. After he did, he fell into a coma for over a year, woke up paralyzed, and then died years later from complications related to the disease.

Rat lungworms have been called "brain worms" because when they enter the human body, they get confused. They can then travel to your brain cavity where they then get stuck. It causes inflammation in the brain leading to serious issues and death.

Snails with this parasite have been found in Hawaii and in many southern US states. Cases are continuing to spread northward as the globe continues to warm. The moral of this story is don't eat raw slugs or snails. Don't even lick them. Be sure to wash veggies from your garden in case their snail trail has left remnants of rat lungworm on your plants. Erika Engelhaupt, the author of *Gory Details*, states that things need to be cooked at 160 degrees for at least 15 seconds to kill any possible parasites.[81] (If you want to read some crazy shit, look up Erika's blog post on rat lungworm.)

Killing snails and slugs in your garden and then leaving them is not a good idea. Even if you kill them, another animal may eat the dead version, continuing the spread of the parasite. So dispose of them after killing them. You can't tell by looking at a snail if they are infested so it is better to play it safe. Remember, they aren't actually pests and if you keep your pets and kids away from them, they can be beneficial for the ecosystem.

People have been eating snails for a long time, and as long as they are properly cooked, they are safe to eat. In France, they eat millions of snails, which they call "escargot," a year.

Their shells have also been used for jewelry, being present in the oldest-known necklace made by humans some 140,000 years ago.[82] And some countries or groups of people once used shells as currency.

Skin care companies have started to use snails and slugs in cosmetics. Their secretions are in face creams and cleansers. The process is currently labor intensive and costly. Also, scientists have found that the quality of the secretions gathered is highly variable across species and laboratories. The quality can be affected by what these creatures eat, how much stress they are under, and if they've had secretions collected too recently. Scientists are trying to create a synthetic version so that keeping large numbers of animals would not be necessary. A lab-made version would then be more readily available to other markets, like wound care and eye drops.

81 Erika Engelhaupt, "Why You Shouldn't Eat a Slug (In Case You Need Reasons)," *National Geographic*, https://www.nationalgeographic.com/science/article/dont-eat-slugs-snails-rat-lungworm-brain-parasite-health-science.

82 David Kindy, "Are These Snail Shells the World's Oldest Known Beads?", *Smithsonian Magazine*, September 27, 2021, https://www.smithsonianmag.com/smart-news/worlds-oldest-jewelry-discovered-in-moroccan-cave-180978766/.

How Human Culture Is Shaped by Snails and Slugs

Snails and slugs show up in many aspects of culture, from creation stories to kids' TV shows to classic literature.

In Diné myth, an important figure in the success of the First People was a figure called Snail Girl. They needed water, so she went to get some with a large bottle on her back. After getting the water, it spilled as she was walking home. This created the first river. They declared that from that day on, the snail would always carry water in the bottle on her back.

Many people associate snails with fertility, sexual desire, the circle of life, the sun, the moon, etc. Christians were the first people to give them negative attributes. They associated the snail with sloth and laziness, some of the worst sins. Snails also appear in Aesop's fables and Shakespeare's plays and in the paintings of Matisse and Dali.

A current artist, Stefan Siverud, has been painting snail shells as a way to make sure they don't get stepped on; he uses non-toxic paint.

Snails and slugs were even integral to the creation of the official royal purple dye that has been used by the aristocracy since the Bronze Age.

You can also find these creatures as characters in *SpongeBob Squarepants* (his pet snail, George) and *Adventure Time*, as well as the mascot of UC Santa Cruz, represented by the banana slug.

Quick Facts: Snails and Slugs

- Snails and slugs are not arthropods but are in the phylum Mollusca, with others like bivalves and nautiluses.

- Do not lick them or eat them raw, as they may give you a rare and potentially deadly disease called rat lungworm. Seriously, cook them very well if you need to eat them.

- Many species have both male and female genitalia and can have some crazy-looking sex.

- People may view them as pests, but they are important to the health of gardens.

- A specific family of marine snails (family Muricidae) is needed to create the royal purple dye worn by aristocracy for centuries.

Chapter 20: Hammerhead Worms and Flatworms (Order Tricladida) (Also Called Planarians)

*T*he variety of animals in this order is impressive; it includes species that are aquatic (both in sea water and freshwater) as well as terrestrial.

While all species share a similar body shape, their overall appearance can be drastically different than their cousins.

Hammerhead worms in particular are alien-looking. These dudes look crazy. Many of them are non-native travelers who showed up in plants or soil transported from afar. Despite their looks, hammerhead forms aren't related to slugs or worms. With their hammerhead, these creatures look quite intimidating, especially compared to the look of many freshwater flatworms that can be found in streams and rivers.

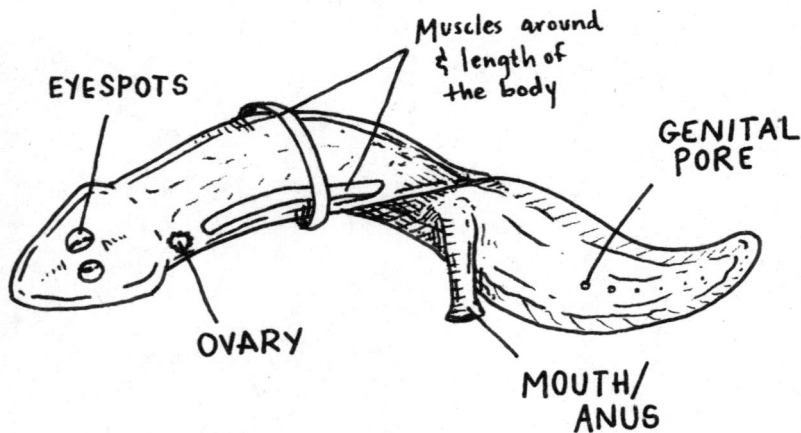

EYESPOTS

Muscles around & length of the body

GENITAL PORE

OVARY

MOUTH/ ANUS

Commonly found in freshwater, flatworms are fairly small and have two visible eyes on the top of their head. They are often used in classrooms to discuss regeneration and other biological experiments. All 20,000 species have a pretty similar body structure to their ancestors, who have been around for about 500 million years—before scorpions![83]

How Hammerhead Worms and Flatworms Shape Our World

All species in this order are predatory. Depending on where they live, they are known to eat earthworms, snails, and other invertebrates. Hammerhead worms will completely cover the body of their prey to prevent them from escaping. All flatworms are basically a giant muscle, which they can use to their advantage when hunting. Terrestrial species also secrete a sticky substance that adds to their effectiveness as predators.

Introduced species are still being studied to determine their overall effect on local ecology. Hammerhead worms are only native to Asia but can now be found in North America and Europe. We will need to wait until more research has been done on them to determine their effect on the food web and other aspects of ecology.

Many non-native species were transferred to different countries via plant soils. As for native species, there are native flatworms in all countries around the world. Some species are food for other animals, whereas non-native species like hammerhead worms do not have any known local predators. There are so many species in this order, often hidden, so it is difficult to determine a working theory on their positive or negative impact on the environment.

---

83 "Flatworms Fact Sheet," *Shape of Life*, n. d., https://www.shapeoflife.org/factsheet/flatworms-ancient-body-plan.

How Humans Can Interact with Hammerhead Worms and Flatworms

Many wildlife biologists advise people to kill hammerhead worms if they are seen, but that is not the recommendation for all flatworms. Aquatic species are often an important bioindicator for water health.

If you do see a hammerhead and plan to kill it, it isn't as easy as you might think. Do not try to smash it or cut it into pieces, because they can regenerate into more worms. The only way to kill them is to throw them in a bag with salt. Some species may cause irritation if handled, so I recommend using gloves or another object to pick them up.

In general, human interactions with this order are positive, innocuous, and often unnoticed. Some people may think they look like leeches, but their body shape, size, and movement are very different. They have no body cavity or circulatory system. Their bodies have one hole used for breathing and expelling waste. They use cilia—little strong feet, like what you see on the bottom of a starfish—to move smoothly and quickly through their environment. They reproduce sexually or asexually and are hermaphroditic.

How Human Culture Is Shaped by Hammerhead Worms and Flatworms

The regenerative abilities of flatworm species have been fascinating to humans for a long time. They are a staple in biological studies and experiments, as well as historical texts and images.

In the 1600s, Japanese wood block prints show hammerhead worms.

American and European scientists became interested in flatworms in the eighteenth century. A Scottish naturalist named Sir John Graham Dalyell eloquently stated that these creatures are "almost immortal under the edge of the knife."[84]

---

84 Sarah A. Elliott and Alejandro Sánchez Alvarado, "The History and Enduring Contributions of Planarians to the Study of Animal Regeneration," *Developmental Biology 2*,

Similar to fruit flies, these invertebrates have had a large influence on biological research that is essential to advancements in human health and disease. They have been studied to see how regeneration of limbs works via stem cells. Research continues to try and determine what we can learn from them and their regenerative qualities. The goal is to determine why certain creatures regenerate, while animals, like humans, do not have the ability to do so.

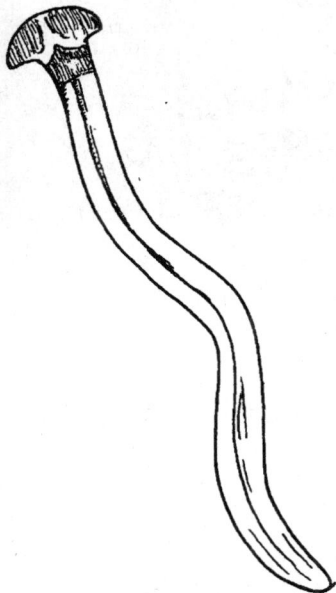

Also not an alien. It's hard to tell at this point.

In one of my favorite shows, *The X-Files*, there is an episode that traumatized me for a while and made me scared to use Johnny-on-the-spots for a long time. The main character in that episode is a human/flatworm hybrid with an exaggerated example of their mouth. Like many other appearances in *X-Files*, the bugs' behaviors were presented inaccurately or in dramatic ways. Flatworms do not bite humans, nor do they lay eggs in you through a bite. The episode explained that toxic waste was to blame for the new species, but there is no such example in the real world. Not even close.

In another *X-Files* episode, a doctor tries to regrow limbs using lizard DNA. They can regenerate their tales like planarians can regenerate their body. In this episode, the doctor is able to regrow a lizard limb on a human. Not possible and not necessarily something researchers are trying to figure out in current flatworm research.

no. 3 (2012), https://doi.org/10.1002/wdev.82.

QUICK FACTS: HAMMERHEAD WORMS

- They may be invasive, but they have been in the US for at least a century.
- If you see one, in order to kill it, you need to put it in a bag with salt. Chopping it up will lead to it regenerating and then turning into many new ones.
- Don't touch them with bare hands, as they may release a toxic chemical that can be irritating to human skin.
- Scientists still don't know the full extent of their influence on ecosystems.
- They have been known to eat earthworms, but the overall effect of them on other bug numbers is unknown.

CHAPTER 21: EARTHWORMS (PHYLUM ANNELIDA)

*T*he creepy crawlies, nightcrawlers, earthworms—whatever you call them, they are all hermaphroditic, and like their main predators (hammerhead worms), some species can regenerate if cut in half.

In order to have sex, the two worms align their sexual organs, then cover themselves with a thick secretion that holds them together. When finished, they lay little cocoons where the babies hang out until they are ready to hatch. The cocoons are a decent size and easily visible. Some are rounded, while others kind of look like cocktail onions.

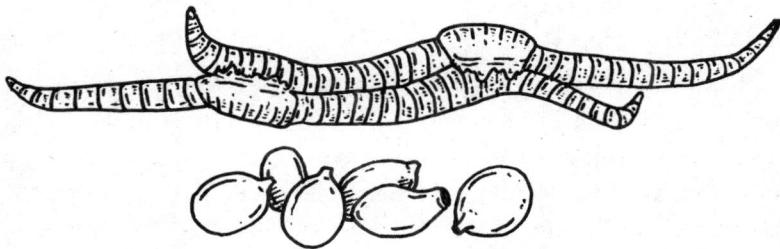

Don't yuck someone's yum.

Of the species in the US, an estimated 33 percent of them (60 species) are introduced species. In the United States, many worms were brought over from Europe because of a decline in native earthworm populations.

Regardless of whether or not they are introduced, earthworms are an essential part of the ecosystem and without them, we would be lost. And they've played an important role in the land for a long time. Earthworm ancestors have been around for 500 million years, whereas their more closely related fossils date back 200 million years.[85] So what would our world look like without them?

How Earthworms Shape Our World

Charles Darwin once wrote about worms: "It may be doubted whether there are many other animals which have played such an important part in the history of the world, as have these lowly organized creatures."[86]

Earthworms are important for soil health, by aerating it and mixing it up, making them a friend to farmers.

People will also purposefully include worms in their compost, called vermicomposting, which breaks down the material faster than regular composting.

Earthworms also serve as food for many animals.

Without them, we as humans would have unusable soil, which would mean inconsistent and bad food production. And we kinda need food.

Researchers recently published a piece on the importance of earthworms in cereal production. They concluded that earthworms in conjunction with other soil biota are essential for around seven to ten percent of all cereal grain production in North America, Europe, the Sahara, and Asia. In sub-Saharan Africa, that accounts for forty million metric tons of grain.[87]

How Humans Can Interact with Earthworms

I remember the first time I went fishing (it was also my last time), my dad bought a large container of nightcrawlers. I love worms so I was more excited to play with them than fish. Then my dad told me I had to pull the worm in half and stick it on the hook. That was not possible

85 Frank Anderson and Samuel James, "The Evolution of Earthworms," BMC Blog Network, June 1, 2017, https://blogs.biomedcentral.com/bmcseriesblog/2017/06/01/the-evolution-of-earthworms/.

86 Charles Darwin, *The Formation of Vegetable Mould, through the Action of Worms, with Observations on Their Habits* (vol. 37), (J. Murray, 1892).

87 Steven J. Fonte, Marian Hsieh, and Nathaniel D. Mueller, "Earthworms Contribute Significantly to Global Food Production," *Nature Communications* 14, no. 5713 (2023), https://doi.org/10.1038/s41467-023-41286-7.

for me, so I made my dad do it. Then the fish I caught swallowed the hook too far and died.

While my experience with fishing was unpleasant, many people love fishing, and earthworms are important for this activity. Some use lures while others will still use live worms. Fish love live worms, so it makes sense to use them as bait. (I just feel really sorry for the worms.)

As with most invertebrates, earthworm numbers and diversity are declining. Factors contributing to that may be habitat destruction, climate change, unsustainable management of soil, and invasive species. Bad management of soil, like tillage practices and intensive land use, decimates the soil and plant roots that earthworms use, which then leads to issues with nutrient cycles.

Since worms are experts at breaking down decaying material, the practice of vermicomposting is gaining popularity, especially in individual homes. Imagine having compost but taking it one step further. Or perhaps in a sideways direction? Worms used in vermicomposting are all species that are not native to the US but most likely found in soil here. The most common are red wigglers since they breed quickly and eat a lot. I lived in an apartment and my neighbor had a vermicomposting bin in his living room. The EPA's website has a great website on how to start your very own vermicompost container.[88] They encourage you to buy worms instead of digging them out of your yard.

And the worm-selling business is no joke. Between commercial sales, agricultural products, and overall soil health, earthworms are extremely important ecologically and economically. Researchers have found that planting perennial flower strips in agricultural fields increases the number of worms by 236 percent.[89] These types of

88 "How to Create and Maintain an Indoor Worm Composting Bin," EPA, n. d., https://www.epa.gov/recycle/how-create-and-maintain-indoor-worm-composting-bin
89 Anna Vaupel, Zita Bednar-Konski, Matías Olivera, Nadine Herwig, Bernd Hommel, and Lukas Beule, "Perennial Flower Strips in Agricultural Landscapes Strongly Promote Earthworm Populations," *Sustainable Agriculture* 2, no. 31 (2024), https://doi.org/10.1038/s44264-024-00040-2.

additions to big AG growing programs have also been suggested for pollinator and insect support. So, not only are we helping worms but several other bug species as well! Seems like a win-win.

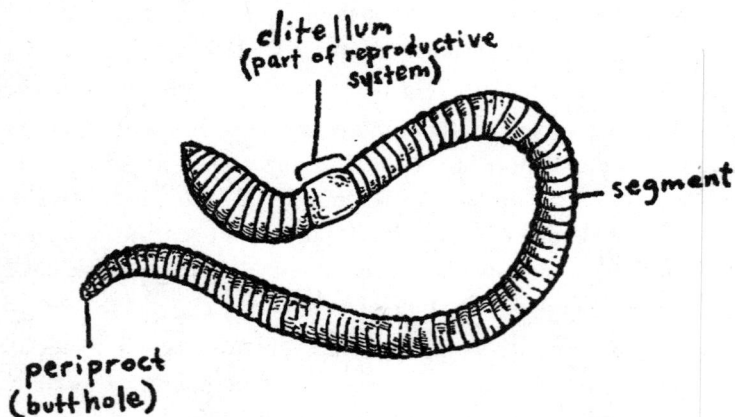

clitellum
(part of reproductive system)

segment

periproct
(butt hole)

How Human Culture Is Shaped by Earthworms

You can find cute representations of earthworms in *Adventure Time*, with Shelby and his brother, or Slimey the Worm, who is Oscar the Grouch's good friend.

Earthworm Jim was a popular character in a 90s video game who wielded a space suit and a blaster to fight evil.

These creatures also appear in songs. *The Book of Scary Stories* that was popular in the late 90s included the song "The Worms Crawl In, The Worms Crawl Out," which is definitely a dark song.

Viagra Boys, a Swedish punk band, has a song called "Worms" about decomposition after death.

While worms are most often associated with food for birds and fish, humans in some areas of the world eat worms. In New Zealand, the Maori have a meal that is considered a delicacy and is made with earthworms. In their mythology, a trickster character transformed himself into a worm to gain everlasting life. Since they have male and female reproductive organs and can regenerate, earthworms were viewed as almost godlike. Today, they are found at the Wild Food Festivals in dishes like worm sushi and crystalized worms.

QUICK FACTS: EARTHWORMS

- Many of the earthworm species you see in the US are introduced/non-native.
- They are essential for soil health and farming would not be possible without them.
- Their numbers are on the decline because of habitat destruction and unsustainable management of soil.
- Some people use worms to compost, called vermicomposting. There are tons of tutorials online to create your own vermicomposting apparatus.
- They are frequently found in media as they relate to decomposition and death, but they can show up in fun ways too like the worms in *Adventure Time* or *Sesame Street*.

CHAPTER 22: PSEUDOSCORPIONS (ORDER PSEUDOSCORPIONES)

*I*f you've ever seen something that looks like a tiny scorpion, with pincers but without a tail, chances are it was a pseudoscorpion. Like spiders, they have the ability to produce silk. They only use it for making small cocoons in which they may mate, molt, or wait out cold weather.

I've never seen one and am so jealous of people who see one in their home. On average, these creatures are about four to five millimeters big, but some can be smaller than a sesame seed or as large as a half inch.

The term "faux scorpion" was coined by the French in the late 1770s.

These bugs live on most continents, including throughout the Americas.

They are known to hitch a ride on other animals as a way to move about. This practice is called phoresy, which is also common in mites. Pseudoscorpions have been observed catching a ride with birds, ants, scorpions, and various other animals. Sometimes, they are so small that they can be found under beetle wings eating mites that have also hitched a ride. Since these guys are so little, it takes a long time for them to get anywhere. Other insects or animals can move great distances, which helps disperse pseudos to new areas.

Let's Jurassic Park this gal.

Since this order is small and delicate, their fossil records are hard to find, but quite a few specimens have been found in amber. In 2024, a new species was discovered trapped in amber in India. The oldest complete specimen dates to 50 million years ago, and based on fossil fragments, scientists believe that these creatures have been around for 380 million years.

How Pseudoscorpions Shape Our World

Pseudoscorpions have weak venom that comes from their mouths, not a stinger. Their mouths are too small to bite humans, and the intended use of their venom is for book lice and house mites instead. Since those two types of invertebrates are often found in homes, pseudoscorpions may be found in human dwellings looking for the buffet. Seeing them in your home isn't bad and it means that they are helping with bug control in your home (and specifically in your bookcase).

Researchers are also looking into pseudoscorpion venom as a possible disinfectant. Antibiotic resistance is becoming more of a problem in hospital settings. In *Bug Life*, I talk about how many hospitals are moving back to using maggots for wound health. Pseudoscorpion venom has a high resistance to MRSA, which is an antibiotic-resistant staph strain. The German team who discovered this states that while this discovery is hopeful, much more testing needs to be done before thinking MRSA is cured. The venom may also have an adverse effect on humans, so the researchers are currently working through that.[90] Fingers crossed that they can figure it out as MRSA is the cause of death for around 10,000 people in the US per year (most of whom are already in the hospital for other reasons).

How Humans Can Interact with Pseudoscorpions

Pseudoscorpions make little nests out of silk, but they don't wait for food to come to their nest. They are active predators who stalk like cats.

90 Pelin Erkoc, Susanne Schiffmann, Thomas Ulshöfer, Marina Henke, Michael Marner, Jonas Krämer, Reinhard Predel, Till F. Schäberle, Sabine Hurka, Ludwig Dersch, Andreas Vilcinskas, Robert Fürst, and Tim Lüddecke, "Determining the Pharmacological Potential and Biological Role of Linear Pseudoscorpion Toxins via Functional Profiling," *iScience* 27, no. 7 (2024), https://doi: 10.1016/j.isci.2024.110209.

The males also use their silk-making capabilities for getting the ladies. They lay down a carpet for the females to follow that leads to their packet of sperm. They also create special silk chambers in which to overwinter.

Many species are solitary, but some live in ant colonies mutualistically. They may also be able to help out honeybees. Scientists are trying to determine if they could be introduced to hives to eat varro mites. Research is ongoing but would be a huge win if a solution to varro mites was discovered. Many scientists agree that the mites have been a large contributing factor to colony collapse.

Recently, a new species of pseudoscorpion was discovered in Boulder, Colorado. It is a cave-dwelling species that lives on bat poop. Scientists named it after the city: *Larca boulderica*. Like many cave-dwelling animals, chances are high that this species of pseudoscorpion does not have eyes.

Caves are always an interesting spot to look for new species and who knows, you may find one the next time you go spelunking. As with all of the other insects I've discussed, you can always whip out your phone and snap a picture for iNaturalist. With bugs that are very small, it is always great to document them when you see them. They are often forgotten or overlooked.

How Human Culture is Shaped by Pseudoscorpions

The first written reference to a pseudoscorpion is in Aristotle's *History of Animals* from around 350 BCE: "Others are like scorpions; they have no tails, and are very small." You may often find them on bookcases and in parchment rolls since they are predators of those insects that eat your books. Believe it or not, they are a library's best friend.

They love libraries as much as the next book nerd.

Quick Facts: Pseudoscorpions

- Unlike scorpions, they do not have a stinger and are incapable of biting humans.
- They have the ability to produce silk and use it to woo the ladies and keep themselves warm during winter.
- They may be useful in helping honeybee hives that have issues with varro mites. Researchers wonder if they can be released in hives and then kill all the varro mites for sustenance.
- Their main courses are insects that eat books, so they are friends to libraries and were even described positively by Aristotle.
- They are very tiny, so may often go unnoticed.

CHAPTER 23: "DADDY LONG LEGS" (HARVESTMEN) (ORDER OPILIONES)

*T*he name "daddy long legs" can be a reference to numerous animals. Some people use the moniker to refer to harvestmen (which this chapter is about) who are not actually spiders. Others use the name to refer to crane flies which are real flies and talked about in my first book. Lastly, some people use the name to describe cellar spiders (which are true spiders).

Harvestmen are arachnids but are in an order of their own. Their body shape is very different from other arachnids. They have a simple oval body with two eyes (spiders often have eight). They also tear off solid chunks of food and eat them, which is very different from spiders, scorpions, and many insects. Most often, they are around two inches but the largest species in Asia can reach upwards of thirteen inches.

The biggest myth surrounding them is that they are highly venomous. *Mythbusters* debunked this in an episode in 2004, and numerous scientists have been trying to reverse this myth publicly for a long time. They do not contain any type of venom or venom glands, nor are they capable of biting humans. Their mouthparts

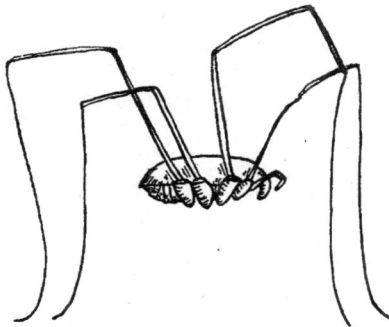

Those chompers are only useful on small prey.

are too small to bite humans. Also, unlike spiders, they are unable to produce silk.

Harvestmen were another favorite of mine as a kid because of their silly legs and the fact that they don't even feel like they are walking on you since their legs are so light.

How Daddy Long Legs Shape Our World

Of all the spider and spider-like beings, daddy long legs may be the most tolerated. They are actually more closely related to mites than true spiders. One of their similarities to mites is that they are omnivorous and known to consume vegetation, both alive and decaying. The remainder of their diet consists of spiders and insects.

When they finish a meal, they will clean their legs using their mouth to remove any particles. Their defenses consist of their odious-smelling liquid secretion, spiky body, running and hiding, and camouflage. Some species may even vibrate rapidly as a threat response. All of these may be surprising to humans but are not dangerous.

There are around 1,820 species worldwide. Unlike most arachnids, harvestmen can be social and have been seen congregating in large numbers. Scientists believe this is a defense mechanism so that they are less likely to be attacked.

If they are attacked and lose a leg (or two even), it has almost no effect on their movement. Some researchers believe that they also congregate to maintain proper hydration levels in individuals. They are a fairly tidy family, and when they have finished feeding, they will clean the web by unhooking the remains of the prey and letting the carcass drop.

They may care for their young, with some species even having paternal care. In some cases where the female of the pair has died, the male will continue protecting the eggs until they hatch. Some have short lifespans, around three months, while some may live up to several years.

How Humans Can Interact with Daddy Long Legs

Daddy long legs don't typically set up shop in houses, as indoor environments are not suitable for them long-term. If you find one in your house, the best suggestion is to relocate them outside. You may find them inside if the weather is bad but finding one is not indicative of an infestation, nor is it indicative that the single one you find is thinking of settling down there.

Last month, in the dead of winter, I felt something brush across my muffin top while sitting on the couch. I thought maybe it was my cat's tail, but it turned out to be a harvestman. Like many bugs in the winter, this one was looking for a place to keep warm. He definitely did not want to live there long-term but happened to be passing through. Unfortunately for him, I placed him back outside, hoping he could find a warm clump of leaves to nestle under. I also find many of them while pulling weeds in the spring, summer, and fall.

How Human Culture Is Shaped by Daddy Long Legs

Salvador Dali included daddy long legs in some of his paintings, including his work "Daddy Longlegs of the Evening - Hope!"

Harvestmen also show up in poetry and other art. Ted Kooser, the US Poet Laureate from 2002 to 2004, has a poem about daddy long legs, with that name.

Fred Astaire was the main role in the musical movie *Daddy Long Legs*, which is the nickname given to him.

A band also bears the name, though their decision for using it is unclear.

At the Pittsburgh Botanic Garden, an artist named David Rogers created a series of bug sculptures. One of the installments was of a twelve-foot-tall daddy long leg that guests can interact with. It is made out of wood, so it kind of looks like a real one, just significantly bigger!

QUICK FACTS: DADDY LONG LEGS

- Daddy Long Legs are not the most venomous spider in the world. They aren't technically spiders, nor do they have any venom that would be harmful to humans. Their mouths are too small to bite humans.
- Unlike many arachnids, they have been known to assemble in large social groups. They may do this as a way to avoid being eaten.
- Some species can live upwards of several years.
- They don't do well in houses, and they will only come inside when the conditions outside are severe.
- Dali included daddy long legs in his art, alongside other bugs like ants and spiders. Read Ted Kooser's poem about them if you have time. It is short and lovely.

CHAPTER 24: VINEGAROONS (ORDER UROPYGI)

V inegaroons are often judged by their appearance. As one headline I saw deftly stated: "creepy to some, dangerous to no one."[91]

At first glance, you realize these creatures are pretty stocky and have a similar shape to a scorpion. They also have fairly large claws and a stinger-looking appendage that sticks out the back. Even though they are sometimes called "whip scorpions," they cannot sting and don't have venom. They are fairly shy and have very few species, at least compared to spiders and scorpions. Simply put, they are all bark and no bite.

Currently, there are only 108 known species. Fossil records put them at about 350 million years old, with little change in structure since then. The term *vinegaroon* comes from their ability to spray a vinegar-like blend from the apparatus at the end of their abdomen (again, not a stinger but a flagellum). Their scientific name comes from the Greek words for "tail rump." They are about 3 to 3.5 inches long.

There are species that live in Arizona, New Mexico, and Texas, though most species live in Asia. Until they reach adulthood, they molt numerous times. Molting takes a few days, whereas fully hardening can take up to four weeks. When they initially molt, they are bright white and darken over the following month.

How Vinegaroons Shape Our World

Vinegaroons live about seven years on average. Most of their time is spent in deep burrows. When they do come out of their burrows to eat, they use their front legs (the non-clawed ones) as what are called sensing legs (or antenniform legs) as a way to tell what is around them.

91 J. Pint, "Meet 2 of Rural Mexico's Insects: Creepy to Some, Dangerous to No One," *Mexico News Daily*, October 15, 2021, https://mexiconewsdaily.com/mexico-living/meet-2-of-rural-mexicos-insects-dangerous-to-no-one/.

They are important to the food web as predator and as prey. As nocturnal hunters, their diet consists of bugs like millipedes, scorpions, isopods, and other arthropods.

Vinegaroons are prey to many vertebrates, such as raccoons, bears, and roadrunners.

For mating, vinegaroons have a complex courtship ritual which can last up to thirteen hours with four different stages. First, they dance with each other for upwards of four hours, during which the male will drag the female around while stroking her with his claws. Second, the male lays a sperm packet on the ground. While it hardens, which can take upwards of two hours, they both remain completely motionless. Once it is hardened, the male will pick it up and put it in the female's reproductive parts. The last part of the mating ritual can take around two hours.[92]

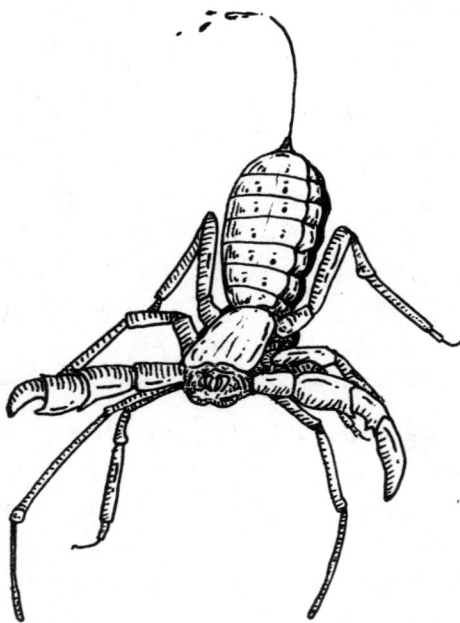

92 "Giant Vinegaroon," Oakland Zoo, n. d., https://www.oaklandzoo.org/animals/giant-vinegaroon.

If the female does not want to mate, she fights the male. The vinegaroon defensive stance is similar to crawfish: claws up, abdomen held high. They also straighten their butt straw.

Once the babies are born, the mother carries them on her back for weeks. She makes sure that they eat and are safe. When the babies leave the burrow for life on their own, in some species the mother dies in her burrow (and is sometimes a meal for her babies) whereas some females go on to live another few years. The toll on the female is significant and leaves her very weak.

How Humans Can Interact with Vinegaroons

As stated above, most species are shy. If you find yourself facing one, best to move along. Maybe snap a picture of it, but it really just wants to get away from you. If you find one in your house (which is extremely unlikely), relocate it outside but watch out for the vinegar it shoots. Those scorpion scoopers might work well.

The chances of you finding one in your house is incredibly slim since human dwellings are not ideal spaces for vinegaroons to thrive/survive in.

Some people do not mind having them in their house since they are good at controlling cockroach and scorpion populations.

They likely won't burrow in your home since they dig deep underground, but they may frequently return for the buffet in your house. Similar to house centipedes, they can help with pests if you feel okay with them being there.

How Human Culture Is Shaped by Vinegaroons

Giant vinegaroons are sometimes called "grampus," which is a phenomenal name.

Vinegaroons appear in the video game *Pocket Ant*, in which these creatures are a mini-boss. In real life, they are definitely a predator of large ants and scorpions, not of humans.

They are part of the exotic pet trade, even though they are native to the US. They are pretty easy to take care of and do not have venom. Their claws can cause quite the pinch however.

In general, scientists try to dissuade people from partaking in the exotic pet trade or taking animals from their native environment to keep in captivity. Their reasoning behind is that (1) we don't know what the species numbers look like so we don't want to add to endangered species; (2) we don't want to release non-native animals into new spaces; and (3) keeping native species and then releasing them can cause its own set of issues like introduction of diseases, parasites, or genetic changes that can negatively impact wild species. These rules are true for all pet species, not just vinegaroons.

Quick Facts: Vinegaroons

- They do not have a stinger, but the appendage sticking out of their abdomen can squirt vinegar, which causes irritation if it gets in your eyes.
- One species lives in the American Southwest, the giant whip scorpion, but most species live in Asia.
- They are very shy and quiet, preferring to keep to themselves.
- People will often keep them as exotic pets. but it is not recommended you do so.
- They live for seven to eight years and partake in extremely long courtship rituals (upwards of twelve hours).

Chapter 25: Whip Spiders (Order Amblypygi)

Similar to vinegaroons, whip spiders most definitely suffer from being judged on their appearance.

I am one of those judgy people because, in my personal opinion, they look terrifying, but in reality they are pretty chill. They almost don't look real. The first time I ever saw one was in a movie and I remember thinking to myself, "What a strange concoction they created for this film." I was very surprised to find out that it was an actual animal. They are also not a spider nor a scorpion, but in an order all their own.

About 260 species have been identified. Fossils date back at least 290 million years. Like vinegaroons and scorpions, whip spiders are nocturnal. Their bodies are extremely flat so that they can squeeze into tiny cracks to hide. When evading predators, they run side to side like a crab. Another attribute they share with vinegaroons is that they have a pair of sensing legs that constantly move around the body. Scientists believe that they use those appendages to find their way home after hunting at night.

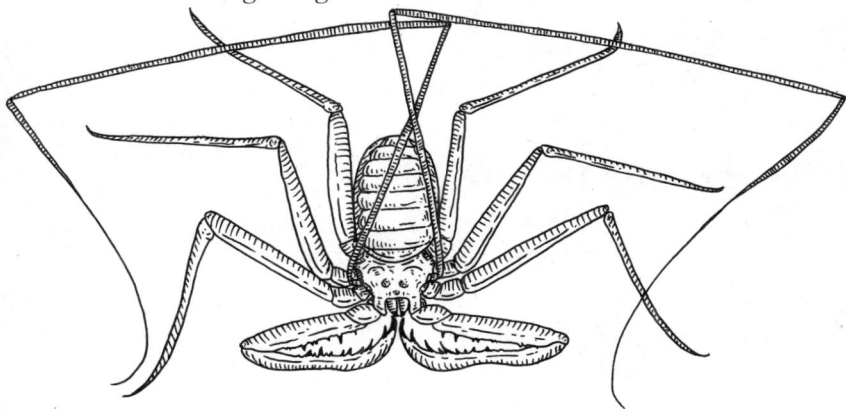

Despite their appearance, whip spiders are delicate creatures.

Their exaggerated pedipalps are raptorial and are covered with spikes which help catch prey when young and help with mating/fighting as adults. They are kind of like mantis arms, laid on their side.

As for their habitat range, the largest number of species live in Asia, but there are species that live in the Americas, Europe, Australia, and Africa. Chances are that, since they are shy and nocturnal, you will only see them at a zoo, even in places where they are native.

How Whip Spiders Shape Our World

Most of their diet consists of arthropods, but they have been known to eat lizards and frogs. They are ambush predators that can move extremely fast. That is also their main defense: running. They don't have venom and their palps are very delicate so not great for fighting larger creatures. Some can regenerate legs. If they lose their palps, they are out of luck. Some species can rub their chelicerae together to make a hiss sound, but most often they prefer to run away.

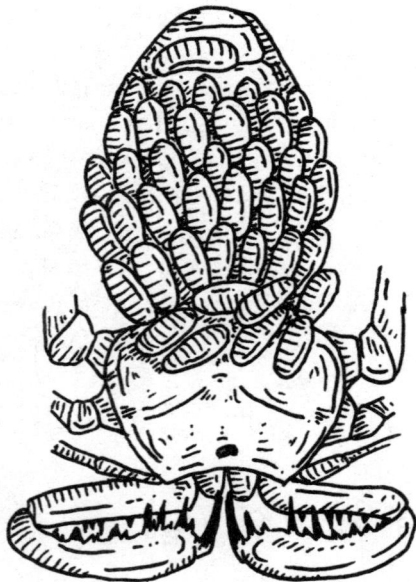

What a loving mother.

Males and females are solitary until it's time to mate. Similar to other arachnids, their mating can be hazardous and drawn out. They will fight if the female does not want to mate. If she does, the male will lay down a sperm sac. After it hardens, the female will be led over the sac and she will place it inside herself. While all of this is happening, they will be caressing each other with their sensing legs. In some species, sexual dimorphism is pretty exaggerated with male pedipalps being almost twice the size of the females; however, in other species there is no noticeable difference between sexes.[93]

When the eggs hatch, the mother carries the young on her back through the first molt. In order to provide a stable space for them to molt, the mother hangs completely motionless, and each young individual climbs to the top of her head, one at a time, to reach their next life stage.

Whip scorpions are unique among arachnids for their parental care. A couple species have been observed caressing their young with their feeling legs. Siblings do similar caressing to each other and stick together through sexual maturity. Once they reach about a year and a half, they no longer spend time together and become aggressive. They all part ways after that. They can live for upwards of ten years.

How Humans Can Interact with Whip Spiders

If you are in northern North America, chances are you will never interact with a whip spider. You are most likely to encounter one in Florida or Mexico. If by some strange event you do find one in your home, you should probably capture it and call your local university entomology department or museum. They may be a non-native species that escaped or a pet that was released. No need to kill it, though catching it might be hard since it is so fast. Again, the chances of this happening are almost definitely zero. Like I said, you'll have to visit a zoo or museum to see one, but you can always ask the zookeepers more questions about them. They are fascinating creatures!

Similar to many of the other bugs I've talked about, there are still many species being discovered each year. This means that the next time you go to the Amazon rainforest or Florida, keep your iNaturalist app handy. You never know what you might find.

93 Iain W. McArthur, Gustavo Silva de Miranda, Michael Seiter, Kenneth James Chapin, "Global Patterns of Sexual Dimorphism in Amblypygi," Zoologischer Anzeiger 273 (2018), https://doi.org/10.1016/j.jcz.2018.02.005.

How Human Culture Is Shaped by Whip Spiders

If you've ever seen the fifth *Harry Potter* movie, a whip spider is what Mad-Eye Moody uses to show the "unforgivable curses"—in that it squeaks when in pain, which is not accurate to their real-life defenses. (Their chelicerae can rub together, but that creates very different noise.)

The inclusion of the tailless whip spider was a creative choice by the filmmakers, as it did not appear in the original text. These creatures do have an intimidating appearance, so when they do appear in popular culture, it is often meant to scare the viewer.

Quick Facts: Whip Spiders

- New species of whip spider are being identified all over the world each year but the majority of known species live in Asia.
- They eat small invertebrates and are then in turn eaten by larger animals like bats and lizards.
- Their large claws are not for defense but to catch prey. When they feel threatened, their response is to run away as quickly as possible.
- Moms are attentive and carry around their young for months until they are ready to go off on their own.
- The fifth *Harry Potter* movie included an inaccurate portrayal of a whip spider: while it was accurate in appearance, it does not make that particular noise in real life.

CHAPTER 26: CAMEL SPIDERS (ORDER SOLIFUGAE)

*C*amel spiders are woefully misunderstood. These little arachnids are neither scorpion nor spider, but in an order of their own. Currently, around one thousand species have been identified, but they are still relatively mysterious to researchers. New species are being found fairly frequently. Scholars, like Dr. Paula Cushing at the Denver Museum of Nature and Science, focus on solifugids and have advanced the field greatly in the last decade. *Solifugae* means "those who flee from the sun," as these creatures are often seen as they run into the shadows.

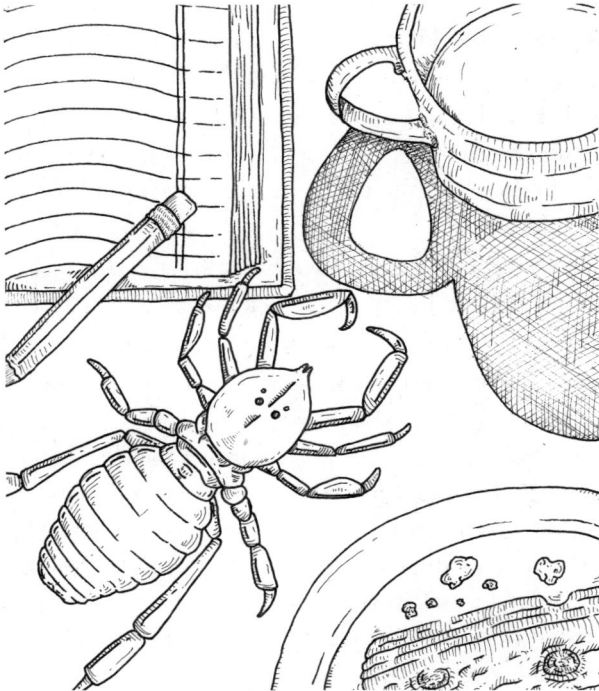

ALFS: Always Looking for Shade

How Camel Spiders Shape Our World

During the Iraq war, videos were showing up on the internet of these behemoth, aggressive spiders that crawled into sleeping bags to bite soldiers. Soldiers would take pictures and purposefully put them in comrades' sleeping bags to scare them. They weren't the first soldiers to use these creatures for entertainment. Soldiers in WWI would make the camel spiders fight, often with the winner eating the loser.

In reality, camel spiders are only about, on average, one to two inches long and have no desire to interact with humans. The largest specimen can be five to six inches long. Yes, that is big, but they don't pose a threat to humans. Their actual prey on the other hand . . .

Their hunting ability is no joke. They don't have venom but their chelicerae are strong and can cause a painful bite. These jaws are very strong and effective at killing.

Camel spiders run after their prey using their back three pairs of legs, and the front pair out ahead of them, sensing as they run. As with many arachnids, their eyes aren't great so their sensing legs and pedipalps help them feel the terrain as they run towards their prey or away from a predator.

If they are defending themselves, they start with a juke, front legs up, chelicerae open, and they don't back down. They chase their prey in open spaces but spend most of their time in their burrow.

As for courtship, they have one set of young per year and reproduce indirectly. The male creates a pack of sperm, flips the female on her back, and uses his jaws to insert the sperm pack into her genital opening. It can appear much more disturbing than that though. Apparently, researchers describe the males masticating (fancy word for chewing) on the female genital opening. They don't know the exact reason but do state that if the male deviates from the norm in any way, she violently responds. This works because the females are much larger than males and may eat them.[94]

94 Jason Bittel, "Camel Spiders Are Fast, Furious, and Horrifically Fascinating," *Smithsonian Magazine*, August 9, 2017, https://www.smithsonianmag.com/science-nature/camel-

Females eat a large amount of food prior to laying eggs since she doesn't eat during the stage in which she guards the eggs. Some stay with their eggs after they've hatched and the young are ready to be out on their own. If they stay in the burrow for too long, they all start to become cannibalistic. Some species lay eggs in those burrows and walk away, never to see their babies. Such is the way of parental care in bugs.

How Humans Can Interact with Camel Spiders

They are known to be aggressive hunters of their prey (i.e., not humans) and only run towards humans because they are trying to stand in your shadow.[95] It may be unsettling since they can run about ten miles per hour, but they can't jump, they don't eat humans, and they don't disembowel camels.

There are species that live in the US, but they are often fairly small. I saw one while living in Arizona and was surprised by how little it was (about half an inch). I was the perfect age to see the Iraq solider online posts about these creatures and believed the "evidence" of their abnormally large size. The little one that I saw did run after my shadow, but I couldn't blame it in the Arizona sun. I wish I had a giant's shadow to stand in in that heat, so I obliged him for a while.

side view

Serious weapons that should be avoided at all costs.

Camel spiders can bite if they feel threatened, but they won't run up and bite you. If you find one in your house, which is most likely to happen in drier, desert-like environments, relocate them using a cup and plate. They wouldn't like being in your house long since they create burrows. Houses don't offer them any

spiders-are-fast-furious-and-horrifically-fascinating-180964439/.
95 Jessie Szalay, "Camel Spiders: Facts and Myths," Live Science, December 16, 2014, https://www.livescience.com/40025-camel-spiders-facts.html.

place to burrow. If they can't get out, they will likely die so relocating them is your best option. Not a bad friend to have in the yard.

How Human Culture Is Shaped by Camel Spiders

Their appearance is strange and startling, especially since they are fast runners (up to ten miles per hour). This has led to many stories and urban legends about them and their surprising appearance, almost all of which are untrue. Some legends say that they can run twenty-five miles an hour. As I mentioned, they can run fast but not quite that fast.

Other legends state that they scream while running (they don't), that they lay eggs in camels, (they don't), and that they can paralyze humans in order to eat them while they are still alive (they can't).

In South Africa, Afrikaans colloquially calls them hair- or beard-cutters, as there was an urban legend that they trimmed hair for their nests. This is also false.

None of these myths are remotely true but have been perpetuated for so long that they are difficult to counter.

Recently, a new Spider-Man character called Sun Spider was introduced in the comics. They are a disabled hero who was bitten by a radioactive spider. None of her abilities are linked to solifugids, which isn't particularly surprising given the similarities between Spider-Man and a real spider. Still, a great inclusion for the spider-verse (even though camel spiders aren't spiders).

Quick Facts: Camel Spiders

- Most are significantly smaller than you'd expect. Some being only half an inch.
- None of the mythology surrounding camel spiders is true. They are not a danger to camels or humans, and they may run towards humans to stand in their shadow.
- Even though they don't attack humans, camel spiders are voracious hunters of small invertebrates and have powerful jaws for that purpose.
- Soldiers used to make them fight each other to the death for their amusement.
- The females care for the young, but if they stay around too long, the mom and siblings will become cannibalistic.
- They do not do well in human dwellings and do not want to be in your house. If you find one, try to relocate it outside where it can eat all the bugs before they enter your house. They are a good yard friend to have.

WHERE ARE ALL THE BUGS GOING?

*T*he Anthropocene is causing significant issues for many bug populations. Humans are changing the earth in ways it has never seen before. And it is messing up numerous animals (including humans) in unprecedented ways. Some scientists are crying "Insect Armageddon" while others are being a bit more cautious. Regardless of what people are saying, a number of insect species are rapidly dwindling.

Some species have adapted well to urban spaces, but biodiversity is extremely important, especially with species that have evolved together over millions of years. Peter Wohlleben has a beautiful book called *The Hidden Life of Trees* that really captures the importance of varied life forms and how the smallest microorganisms and the earth as a whole completely rely on each other. While he may not be an entomologist, his holistic look at nature includes the importance and interconnectedness of all living things, including insects. I highly recommend reading it.

Reading about it and talking about it are one thing, but what are we supposed to do in the face of such destruction?

The main recommendations for addressing these declines are to counteract habitat loss, reduce the number of pesticides being used, and address ways to slow climate change.[96] In order to do that, we need to protect habitats, change laws/legislation, and increase education about bugs.

Below are some steps you can take to promote healthy, bug-friendly spaces. I like to call it the "I'm not lazy, I'm just saving the world" approach:

---

96 Matthew L. Forister, Emma M. Pelton, and Scott H. Black, "Declines in Insect Abundance and Diversity: We Know Enough to Act Now," *Conservation Science and Practice* 1, no. 8 (2019), https://doi.org/10.1111/csp2.80.

- *Don't cut the grass too often*: Insects like long grass to hide in and create their nests
- *Don't pull your weeds if they are flowering*: Scientists are torn on this, but pollinators love dandelions and other flowering weeds. If you can't afford to plant anything, it's better than nothing.

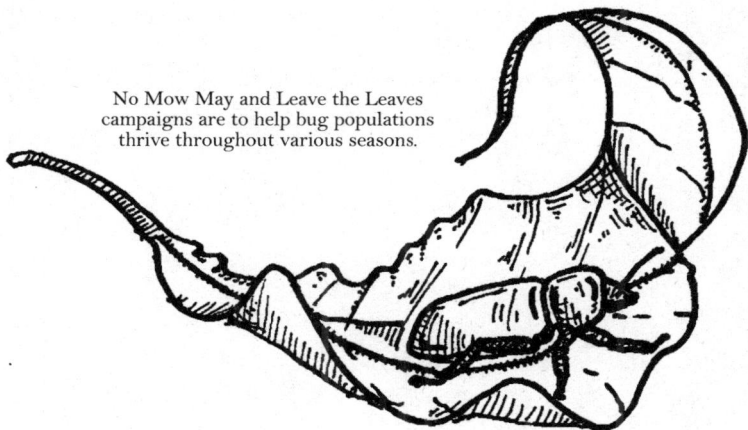

No Mow May and Leave the Leaves campaigns are to help bug populations thrive throughout various seasons.

- *Don't pull down your dead plants or rake your leaves*: Many insects, including solitary bees and moths, will overwinter in dead plants (also dead leaves). Be sure to leave it all there so that they can keep sleeping.
- *Keep your lights off at night*: Many insects can be disoriented by light so keeping them off helps them locate what they actually need to see.
- *Plant native pollinator flowers (if you can)*: This one isn't actually easy since you need space and money, as well as knowledge of what to buy. If you do have space, time, and money, look at the Xerces Society website to determine what is native in your area. Many cities give away free seed packets. Be sure to ask if the packets are free of neonicotinoids.

- *Donate to organizations that address these issues*: Xerces Society for Invertebrate Conservation, the Center for Biological Diversity, and many others care about and fight for all the little creatures that may not get as much love as they deserve: xerces.org/donate and act.biologicaldiversity.org/a/donate-now.

How Do I Keep Bugs Out of My House?

Not a bad question. My main recommendation would be *do not spray pesticides* until you've addressed everything else that might help.

Let's chat about Integrated Pest Management (IPM) for a bit. IPM is the use of more preventative measures and less toxic attempts at ridding your home of bugs. This is important in sensitive environments (i.e., where people live and spend most of their time) since many of the typical pesticide sprays are not as safe for humans as you may think.

Why don't I recommend pesticides? Rarely do they do what we want them to do and then we have a toxic residue that is harmful to you, your family, your pets, and other bugs that you aren't trying to kill.

Many pest control companies say that their spray is safe for humans and animals, but they are full of shit, especially with how they apply it. Pest control businesses, especially national chains and the like, rely on capitalism, and a great way to make money is to tell people they need their house sprayed twice a year. And they also cause concern around bugs that will most likely never cause your home issues (think carpenter ants or American cockroaches). Research shows that baseboard spraying is highly ineffective and that it rarely addresses the issue. More directed, specific application, especially that which only targets a single species, is most effective.[97,98]

---

97 Youn-Hyung Lee, Ho-Hyun Kim, Jeong-Il Lee, Jeong-Hee Lee, Hee Kang, and Jae-Young Lee, "Indoor Contamination from Pesticides Used for Outdoor Insect Control," *Science of the Total Environment* 625 (2018), https://doi: 10.1016/j.scitotenv.2018.01.010.

98 Islam Md Meftaul, Kadiyala Venkateswarlu, Rajarathnam Dharmarajan, Prasath Annamalai, Mallavarapu Megharaj, "Pesticides in the Urban Environment: A Potential Threat that Knocks at the Door," *Science of the Total Environment* 711, https://doi:10.1016/j.scitotenv.2019.134612.

Without resorting to toxic chemicals, you can greatly reduce the number of bugs you see in your house with the following tips:

Most issues with unwanted bugs in homes can be fixed without sprays.

- *Check and fix any leak*: As was discussed in many chapters, a lot of "pests" like humid, wet, and moldy conditions, so be sure to find and fix any leak early. Removing all standing water is also a good idea. Not bird baths but old tires filled with rainwater.

- *Clean up your food*: Bugs love our food almost as much as we do, so be sure to clean up so that you don't get them all hanging out under your stove.

- *Be sure that your screens and the space between the floorboards is blocked*: They can get through tiny cracks so be sure the space between the wall runner and the floor is closed and that there are no gaps in the window screens. It should reduce the number of insects who randomly stumble in.

- *Be sure any cracks in foundations, walls, or any other crevice that bugs could make their way through are blocked*. Caulk, tape—whatever you want and think will stand the test of time. With tape, you will most likely need to replace that often. Be sure to have something blocking under doors and in doorways.

- *Vacuum regularly*: Many bugs love hair and skin flakes, so vacuum up all that pet fuzz so that they don't eat it.

- *Make sure that there is no vegetation touching the house*: Keep leaves and other decaying matter eighteen inches from the foundation of your house.

- *Change outdoor lightbulbs to orange/yellow LED or yellow "bug lights"*: Regular LED lights can still be attractive for some insects, so choosing yellow is important. Also, turn them off when not needed. This will reduce the number of flying insects who may accidentally make their way indoors through complete confusion. Remember, their brains are really tiny.
- *If you do see a bug, try to figure out what it is before killing it*: Chances are what you see is harmless, so don't immediately fear it. Approach it with curiosity and respect.
- *If the bugs in question continue to be a serious pest, knowing what you have is important.* General sprays and traps don't necessarily kill or address what you need to rid your home of a pest. Knowing what species are in your house is extremely important. Do not ask chatGPT what they are but instead ask a local scientist or expert (preferably not large, well-known national pest companies, since they'll most likely try to sell you a spray plan).
- *Set out sticky traps, insect specific, to determine what type of infestation you may have or if you have an infestation at all.* If you only regularly catch non-problematic species like crickets and millipedes, you probably have a crack or leak somewhere you haven't found yet. Insect-specific traps have less intense glue so you won't catch larger animals. Personally, I think sticky traps are seriously problematic, but you do you.
- *Sprays and dusts should be your last resort and done in very specific places to avoid toxicity issues.* Do not apply to anything that can be touched by humans or pets, like rugs, bedding, couches, etc. Spraying baseboards is not effective so only applying spray to cracks that are not able to be closed is the best suggestion.

Do not spray pesticides in your home unless it is absolutely necessary. When we had bedbugs, they sprayed so many times, but it turns out that doesn't do anything really for getting rid of bedbugs. So not only did it not address the issue, but it also exposed us to an intense pesticide that was sprayed in our tiny apartment on four separate occasions, including on our mattress and couch. I wish I knew then what I know now.[99]

---

99 The process for removing bedbugs is completely different from most. Insecticides are useless. Heat treatment is the best option for eradicating them.

Conclusion: Now What?

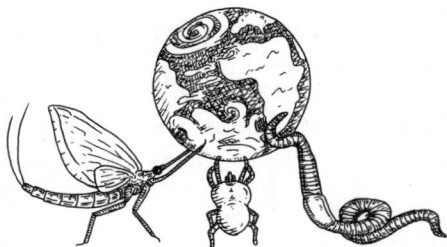

*A*s I write this, in the US, the administration that got into office in January 2025 is making drastic cuts to programs that help address issues related to endangered species.

Many species that people know of or care about are large animals which are called charismatic megafauna (grizzly bears, elk, etc.), but the smaller, less-seen animals are desperately in need of our protection.

Not everyone may notice when they're gone, but if they all disappear, we are fucked. Truly fucked. If you like drinkable water and edible food variety, those things are impossible without invertebrates. Without mussels and with increases in pollution, our water will be undrinkable. Without aquatic invertebrates, other animal populations who are integral to the food web will start to disappear. Without the tiny creepy crawlies that eat detritus, we will be buried in garbage and lose the base of the food web. Without bugs and with how we mass farm, our soil will not have essential nutrients.

Bugs touch every aspect of our lives whether we realize it, want to accept it, or prefer to ignore it. Whichever way you slice it, without invertebrates, everything falls apart.

So let's do all we can to make sure that doesn't happen.

FURTHER READING

Rodolfo Dirzo, Hillary S. Young, Mauro Galetti, Gerardo Ceballos, Nick J. B. Isaac, Ben Collen, "Defaunation in the Anthropocene," *Science*, 345, no. 6195 (2014), https://doi:10.1126/science.1251817.

Matthew L. Forister, Emma M. Pelton, and Scott H. Black, "Declines in Insect Abundance and Diversity: We Know Enough to Act Now," *Conservation Science and Practice* 1, no. 8 (2019), https://doi.org/10.1111/csp2.80.

Yuya Fukano and Masashi Soga, "Why Do So Many Modern People Hate Insects? The Urbanization–Disgust Hypothesis, *Science of the Total Environment* 777 (2021), https://doi.org/10.1016/j.scitotenv.2021.146229.

Caspar A. Hallmann, Martin Sorg, Eelke Jongejans, Henk Siepel, Nick Hofland, Heinz Schwan, Werner Stenmans, Andreas Müller, Hubert Sumser, Thomas Hörren, Dave Goulson, Hans de Kroon, "More than 75 Percent Decline over 27 Years in Total Flying Insect Biomass in Protected Areas," *PloS One* 12, no. 10) (2017), https://DOI: 10.1371/journal.pone.0185809.

Forest Isbell, Patricia Balvanera, Akira S. Mori, Jin Sheng He, James M. Bullock, Ganga Ram Regmi, . . . and Meredith S. Palmer, "Expert Perspectives on Global Biodiversity Loss and Its Drivers and Impacts on People," *Frontiers in Ecology and the Environment* 21, no. 2 (2023).

Julia Janicki, Gloria Dickie, Simon Scarr, and Jitesh Chowdhury, "The Collapse of Insects," Reuters Graphics, December 6, 2022, https://www.reuters.com/graphics/global-environment/insect-apocalypse/egpbykdxjvq/.

Alison McAfee, "The Problem with Honey Bees," *Scientific American*, November 4, 2020, https://www.scientificamerican.com/article/the-problem-with-honey-bees/.

Christine Peterson, " 'Like Butter for Bears': The Grizzlies who Dine on 40,000 Moths a Day," *The Guardian*, https://www.theguardian.com/environment/2023/oct/08/grizzly-bears-eat-moths-wyoming.

Vincent H. Resh and Ring T. Cardé, eds., *Encyclopedia of Insects*, 2nd edition (Academic Press, 2009).

Francesco Ricci, "Social Implications of Malaria and Their Relationships with Poverty," *Mediterranean Journal of Hematology and Infectious Diseases* 4, vol. 1 (2012) https://doi.org/10.4084/MJHID.2012.048.

Chris D. Thomas, T. Hefin Jones, Sue E. Hartley, " 'Insectageddon': A Call for More Robust Data and Rigorous Analyses," *Global Change Biology* 25, no. 6 (2019), https://doi:10.1111/gcb.14608

"Understanding Neonicotinoids," Xerces Society, n.d., https://www.xerces.org/pesticides/understanding-neonicotinoids.

More Rad Titles from Karyn Light-Gibson

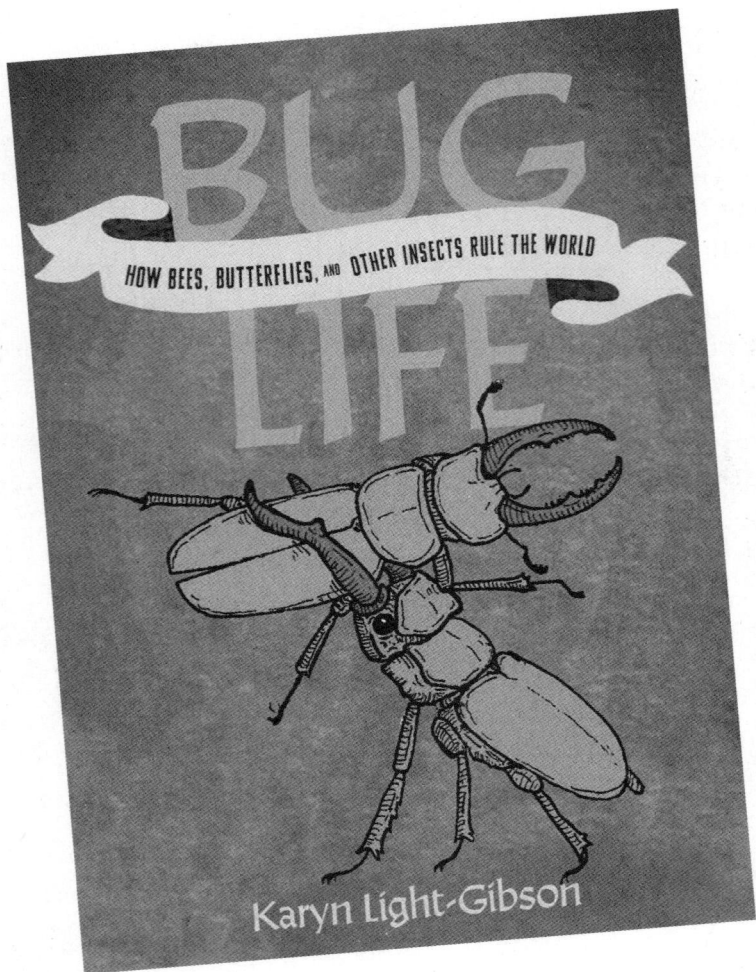

BUG LIFE

HOW BEES, BUTTERFLIES, AND OTHER INSECTS RULE THE WORLD

Karyn Light-Gibson

THE SEX LIVES

HOW BEES, SPIDERS, WORMS, MITES, AND OTHER INSECTS GET IT ON

OF BUGS

Karyn Light-Gibson

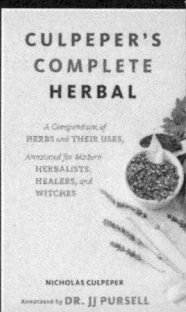